Process Plant Construction

Process Plant Construction

A Handbook for Quality Management

Patrick J. Noble

WILEY-BLACKWELL
A John Wiley & Sons, Ltd., Publication

This edition first published 2009

Blackwell Publishing was acquired by John Wiley & Sons in February 2007. Blackwell's publishing programme has been merged with Wiley's global Scientific, Technical, and Medical business to form Wiley-Blackwell.

Registered office
John Wiley & Sons Ltd, The Atrium, Southern Gate, Chichester, West Sussex, PO19 8SQ, United Kingdom

Editorial offices
9600 Garsington Road, Oxford, OX4 2DQ, United Kingdom
350 Main Street, Malden, MA 02148-5020, USA

For details of our global editorial offices, for customer services and for information about how to apply for permission to reuse the copyright material in this book please see our website at www.wiley.com/wiley-blackwell.

Library of Congress Cataloging-in-Publication Data
Noble, Patrick J.
Process plant construction : a handbook for quality management / Patrick J. Noble.
p. cm.
Includes bibliographical references and index.
ISBN 978-1-4051-8725-1 (hardback : alk. paper) 1. Factories–Design and construction–Quality control. 2. Building–Superintendence. I. Title.

TH4511.N63 2008
690'.54—dc22

2008022790

A catalogue record for this book is available from the British Library.

Set in 10.5/13 pt Trump Mediaeval by Newgen Imaging Systems Pvt. Ltd, Chennai, India
Printed in Singapore by Utopia Press Pte Ltd

1 2009

Contents

Preface

Head office engineering of process plant projects, together with the associated procurement of materiel are, generally speaking, well structured within a framework of formal procedures. Unfortunately this is not always the case concerning quality management of the construction phase, for all too often serious thought is only given to it once the site is open, contracts are placed and work has started. Under these circumstances it may be possible to create a rudimentary system of quality management making the best of a bad job, but such an approach is to say the least unsatisfactory.

My contribution to improving this situation is to present in this handbook a straightforward, commonsense model of quality management, based on my own experience, which can be readily adapted to the needs of any process plant project. I trust that it will:

- Prove to be a useful companion and guide to those persons directly involved, in particular the younger technicians and engineers, providing them with a 'user-friendly toolkit' to create and operate effective construction quality management systems
- Also encourage top management of the various entities concerned to examine their culture and methods and, if necessary, bring about change, so that quality management of construction is a central facet of corporate and project strategy and not just an afterthought

Although this handbook is focussed on quality management systems for the construction of process plants, most of the principles and procedures proposed can be applied to other types of project.

Patrick J. Noble

Acknowledgements

I wish to express my sincere thanks to the following persons for their help and encouragement in creating this handbook: Karim Benziane, Robert Broatch, Philip Broos, Chris Fox, Patrick Hehenberger, Jacques Rozenwajn and last but not least my wife, Gilberte.

I am also grateful to the Institution of Engineering and Technology for permission to base Figures 1.1, 1.2, 4.2 and 11.7 on figures in my article, 'Quality Assurance in Construction of Process Plants', published in *Engineering Management Journal*, Vol. 7, no. 4.

My thanks go also to Wiley-Blackwell, in particular to Madeleine Metcalfe, James Sowden and Lucy Alexander, as well as the team at Newgen Imaging Systems, for their very professional work in producing this handbook.

Introduction

Through the 11 chapters of the handbook, the reader is introduced progressively into a methodology for mastering the management of quality on a process plant construction site. The various principles presented are set into context by the inclusion of a number of case studies.

Chapter 1 shows the pivotal role occupied by the construction phase of a project, at the same time defining the term 'quality'. It goes on to identify the different parties concerned and the relationships between them. Finally, the necessity of defining right from the outset project construction strategy is emphasised.

Chapter 2 defines the concept of quality assurance, how it has been codified in the ISO 9000 series of standards and how the latter can be translated into a meaningful quality management system on a construction site. A negative image of ISO 9000 is not uncommon in the construction industry; this is explained and an appropriate response is given.

Chapter 3 focuses on the site and the role of the various actors present, whose relationships are illustrated in system terms. The sharing of responsibilities for attaining quality is presented in the form of a matrix. Then plant extensions and modifications are discussed as well as adapting quality management principles to small construction sites.

Chapter 4 discusses the central role of the engineering, procurement and construction management (EPCM) contractor on the construction site, its organization and the duties of key members of its staff. An approach is given for the preparation from scratch of a quality plan for this organization and a typical organization chart is included.

Chapter 5 concentrates on the part played by the various specialist construction contractors and how they can be selected. Guidelines are set out for the initial site meeting and then models are proposed for the various documents to define and then record the required programme of inspection and tests.

Chapter 6 is concerned with the interface between the head office engineering design function and the construction site. The different types of documents concerned are defined, as well as the document control function to ensure that the right ones are available to the persons who need them. A typical site-generated engineering query resolution system is proposed and the importance of establishing 'as-built' documents is stressed. The chapter closes with a discussion on the role of information technology to facilitate management, storage, transmission of and access to documents.

Chapter 7 covers the site management of materiel, starting with the different sources of supply and underlining the importance of effective upstream engineering, procurement and at-source inspection to avoid problems on site. Incoming inspection and ongoing maintenance are explained with model forms to manage these activities. The chapter concludes with a discussion on traceability and spare parts.

Chapters 8 and 9 discuss, respectively, management of nonconformities and quality audits, proposing model procedures complete with sets of forms. These two activities are vital, providing means of monitoring the quality management system to detect and correct anomalies, but they can generate resentment. An approach is proposed to avoid this situation and bring about cooperation.

Chapter 10 cites another management tool, that of periodic reviews of the site quality management system, giving a model list of areas of investigation. The necessity of a final site project report is underlined to enable lessons learnt to be fed into the corporate pool of experience and know-how for the benefit of future projects.

Chapter 11 is concerned with the formal completion of construction, leading on to the watershed milestone of turnover to the owner. The various stages are defined with model procedures and forms to complete and document the transfer.

There are ten appendices (A to J), each devoted to a specific discipline, starting with civil works and buildings, then progressing through mechanical equipment erection, electrical installation, and so on, terminating with painting and thermal insulation. The order of the appendices corresponds broadly speaking to the sequence in which the different works are put in hand on site. This part of the handbook is not in any way a technical treatise, for each discipline represents a vast range of knowledge meriting its own handbook. It is instead a management review aimed at the nonspecialist, presenting briefly for each discipline a description of the works likely to be met on site, evoking the parties involved and highlighting the quality issues to be addressed. Typical inspection and test programmes are outlined.

However, the definition and the execution on an inspection and test programme for an actual project must be solidly based on the specifications, drawings, standards, procedures and other documents applicable to the project in question, with full account taken of the legal and regulatory requirements concerned, especially with regards to occupational health and safety, as well as the protection of the environment.

Downloadable forms

For those chapters proposing typical procedures with model forms concerning responsibilities on site, supplier site visits, quality plans and records, engineering queries, incoming inspection and maintenance, nonconformities, audits, management reviews and construction completion and turnover, the forms concerned can be downloaded from the Wiley-Blackwell website: www.blackwellpublishing.com/noble. They can serve as a basis for the reader to create his/her own forms adapted to the needs of a specific project.

1

An Overview

1.1 Introduction

A process plant is an industrial facility which transforms chemically and/or physically bulk raw or partly processed material into a useful new product. Such plants are to be found in a wide range of industries, including oil and gas, chemicals, petrochemicals, water supply and food. The process plant, properly speaking, is supported by utilities, off-sites and infrastructure to form a coherent functional whole.

The creation of a new plant or the extension of an existing one necessitates a preliminary design centred on the process. Further engineering develops the initial concept to the point where detailed specifications and drawings can be produced, leading to the procurement and delivery to the site of the necessary materials and equipment. Completion of construction leads on to commissioning and start-up, when the operational robustness of components and systems is demonstrated, as well as the ability of the plant as a whole to perform to specified requirements. Figure 1.1 is a sequential chart for a typical project showing the relationship between construction and the other phases both upstream and downstream.

Within this context, 'quality' of the finished plant is defined as the totality of features, characteristics and attributes that

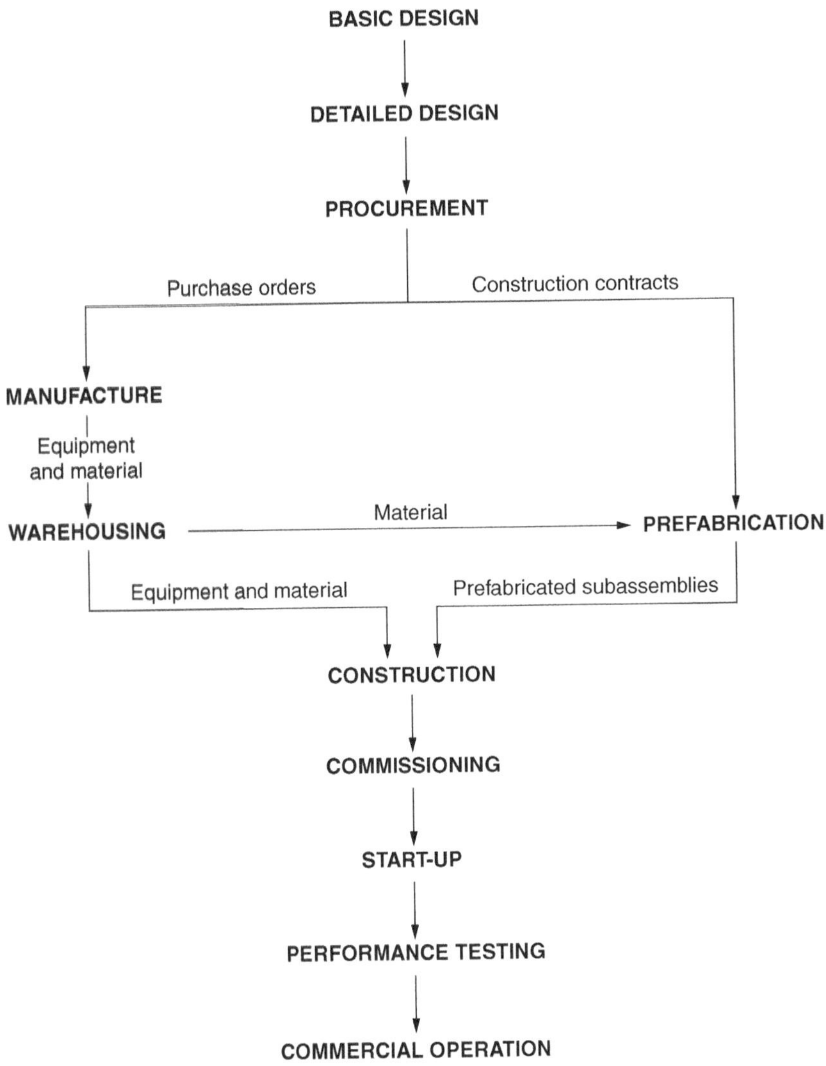

Figure 1.1 Typical process plant project.

enables the plant to satisfy specified requirements. More succinctly, quality can be perceived as 'fitness for intended purpose', thus meeting the expectations of the parties involved. Failure to attain the desired level of quality may be due to root causes present at any stage of the project. These may be conceptual, physical or organizational.

From the viewpoint of quality, the construction phase is the project crossroads, for it is here that design and procurement activities come to fruition as the plant progressively assumes its physical existence and identity.

Also it is here that the foundations are laid for trouble-free commissioning and start-up. Thus, effective quality management during construction makes a key contribution to successful completion of the project.

Whilst this handbook is aimed at quality management of the construction of process plants, the principles, approach and methods proposed are to a large extent relevant to a range of other industrial installations. These include *inter alia* power plants, bulk liquid storage depots, sewage treatment works and waste disposal facilities.

1.2 Material, equipment, materiel and works

These four words appear often in this handbook, so it is important to understand from the outset the meaning given to them. **Material** comprises items not having a specific designation and which could be incorporated into a number of different locations on the site. Low-tension cables and fittings, cable trays, most pipes and fittings as well as building materials come under this category. **Equipment** comprises items which have a designation and usually a tag number, 'Cooling Water Pump N° 1 – P 201' for example, and which are to be installed in a specific functional and/or geographical location. **Materiel** is a collective noun that we use to refer at the same time to both material and equipment. As for **works**, it is a wider concept embracing materiel used and the activities necessary to install, erect or transform it into the whole or part of the plant.

1.3 Interested parties

A number of different parties are involved in the realization of a process plant and they may all influence directly or indirectly the quality of the finished facility. There is, in fact, a range of different possible configurations concerning who finances the project, who initiates and realizes it, who will own the completed plant and commission it and who will run it on a commercial basis once it is operational. Traditionally, the same organization, an oil company for example, could initiate a project, finance it and steer it through the engineering, construction and commissioning phases, and then operate it throughout its working life. If the company concerned possessed sufficient resources in terms of know-how and experienced and qualified manpower, the project could be handled in-house with minimal participation of other parties. More often, an engineering, procurement and construction management contractor would be appointed to be responsible for designing the plant, procuring the materiel and supervising construction, before turning over the plant at a predetermined point after construction was finished.

In recent years while the above cited scenario is still present, a number of alternative project structures have emerged, which require closer collaboration between the different actors, with a redistribution of roles, risks and responsibilities. They appear under a range of appellations such as 'joint venture', 'alliance' or 'partnership'. A detailed analysis of these arrangements is beyond the scope of this handbook, but let us cite hereunder some of the more prevalent examples:

- A horizontal link such as a joint venture between two or more engineering, procurement and construction management contractors for the duration of a project to enable the partners to work from a stronger financial base, to share risks, to pool resources and know-how and to better respond to the expectations of the owner
- A vertical link being an alliance between the owner and the engineering, procurement and construction management contractor integrating project management, simplifying the exchange of information and thus streamlining decision taking
- Extending the vertical liaison downstream to include suppliers of major equipment items, generally on the critical path and with long lead times, so as to inject supplier specialist knowledge early into the design, saving time

by avoiding the bidding process and improving quality by reducing the number of change orders in later stages and even improving constructability and simplifying maintenance of the finished plant
- The ultimate in vertical integration where a consortium finances the plant, remains in ownership of it, designs, builds and commissions it and continues to maintain it throughout its working life, renting it out as an operational facility to a production company

The vertical links cited above generally give rise to a higher initial nominal contract price, but which should be more than compensated by tangible savings downstream as well as intangible benefits. Avoiding costly and time-wasting adversarial situations, unfortunately all too common in the construction industry, has been the main driving force behind the promulgation of these approaches. Results have been positive where mutual trust has been generated and maintained, objectives of the participants have been well defined and top management has provided the necessary inspiration. Failures have occurred where these factors have been lacking.

These innovative styles of managing process plant projects can be accompanied by a number of financing structures, for example:

- Partners agree a target cost to be compared with ultimate actual cost, sharing the eventual savings or cost overrun, thus providing motivation to the parties concerned.
- The plant remains the property of a bank, which has financed the project, to be handed over on a sale/lease-back arrangement to the company which has initiated the project, driven it through to commissioning and start-up and will take it forward into commercial production.
- In certain situations the addition of regional, national or international public money to private investment as part of government policy will be made to encourage industry and employment.

Whilst the introduction of these novel contractual and financial structures generally revolutionizes many aspects of running a project, in fact they make little difference to the day-to-day management of quality on site. Whatever the arrangement chosen, there will always be specialist contractors who do the work, there will always be an organization, whatever it may be called, which manages, coordinates and supervises everything which happens on the site, and there will always be an entity to whom the completed plant is turned over. The only effective difference between one project and another will be the identities and names of the parties fulfilling these roles. The essence of the relationships between them, as far as construction quality management is concerned, will remain substantially the same. So for the sake of simplicity and in order to give this handbook focus and coherence, we have retained a single traditional model, for which we have defined hereunder the appellations and roles of the parties concerned. However, the methodology is valid whatever the financial, contractual and management structure, and so transposing the ideas proposed in this handbook into a form adapted to a specific project founded on another commercial/contractual model should not present too much difficulty.

In our chosen model the **owner** is the entity which initiates the project, finances it, carries out the basic design and which will take possession of it and operate it once it is complete. Let us assume that this organization does not possess engineering, procurement and construction departments of sufficient size and capability to run the project in-house. Consequently, the owner engages an **engineering, procurement and construction management (EPCM) contractor**.

Suppliers under the supervision of the EPCM contractor furnish the necessary materiel for the project. The work on site will be carried out by **construction contractors** who are also under the supervision of the EPCM contractor. **Special service providers** may be engaged by the owner, by the EPCM contractor or by a construction contractor for duties such as surveying, civil site and laboratory testing, for carrying out destructive tests and nondestructive examination associated with welding or for the interpretation of results obtained from these activities.

Finally, there are outside organizations whose exigencies have to be met if the project is to advance. There are, first of all, the **utility companies** providing electricity, gas, water, telecommunication facilities and sometimes steam. They may be privately owned, nationalized industries or government departments. There is also the **insurer** of the project who will have requirements to be met. Finally, there are the **authorities** responsible for enforcing the applicable legislation and dependent regulations in the country in which the plant is to be built and also possibly in the country or countries of manufacture of materiel incorporated into the works.

These appellations of the parties involved and theirs roles as defined above will be assumed throughout this handbook. Whilst the perspective presented is essentially that of the EPCM contractor, the same principles can be adapted to the situation of the other participants. In Section 3.12 guidelines are given for applying the principles and procedures proposed in this handbook to the smaller construction site, where structures and facilities are present on a smaller scale.

Concerning the process technology, it may be owned or licensed by the owner or by the EPCM contractor. For the sake of simplicity we have assumed that this technology is owned by the owner and embedded in the basic design.

1.4 Project strategy

Very early in the project, fundamental decisions have to be taken concerning the manner in which the totality of activities and their corresponding responsibilities are to be divided into lots. The sort of questions to be addressed are as follows:

- What are the depth and scope of the preliminary design to be provided by the owner to the EPCM contractor?
- Will the EPCM contractor carry out all detailed design across all disciplines, or in certain specialized areas will this be part of the scope of work of the selected construction contractor, in the case of, for example, a complete water treatment plant or an electric substation?
- What materiel will be freely issued to the construction contractors and what materiel will they be expected to

procure? Taking the piping discipline for example, will valves, pipes and fittings be freely issued to the construction contractor or will the latter be required to supply them?
- To what extent will procurement of materiel be grouped? Will, for example, valves, pipes and fittings be procured through a multiplicity of purchase orders awarded directly to a large number of suppliers, or alternatively will this procurement exercise be devolved to a single or a small number of stockists?
- For each discipline will there be a single site-wide contract or will there be a split by geographical/process zone to form two or more contracts?
- Will related disciplines be separated contractually or combined? For example, will mechanical equipment erection and piping prefabrication/erection be the subject of two distinct contracts or of a combined contract, which in effect shifts the interface between the two disciplines from the EPCM contractor to the single chosen construction contractor?
- Very importantly, what will be the responsibility interfaces between the owner and the EPCM contractor towards project completion and how will they evolve over time? What are the definitions of 'construction completion', 'commissioning', 'start-up', 'performance testing' and 'commercial operation'? Which organization will be responsible for each of these phases and how will the work scope be shared?

These decisions are often perceived simply as a commercial, planning and administrative matter, but they are in fact also very much a quality issue, as they determine interfaces between the interested parties. In tangible terms, the outcome of the above exercise is a list of potential purchase orders and construction contracts as well as tables defining the scope of responsibilities between the owner and the EPCM contractor. These are key control documents to be completed with more detail as the project advances, in particular, with the names of the appointed suppliers and construction contractors. To get a global image of this exercise of determining project strategy, let us look at Figure 1.2, where the diagram shows notionally

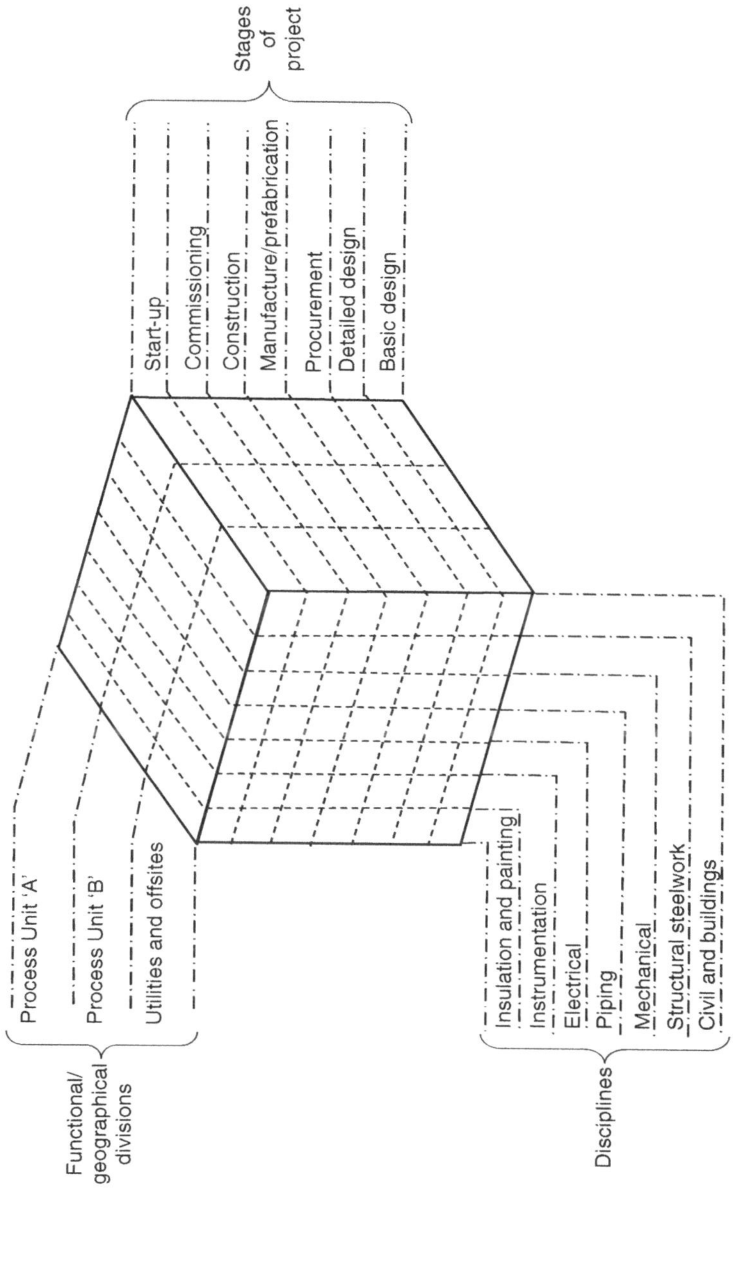

Figure 1.2 Project strategy.

the totality of a project represented as a cube, able to be sliced in three planes mutually at right angles representing:

- The successive stages of the project from preliminary design through to completion
- The various disciplines involved, which on a site are implemented broadly speaking in sequence starting with civil works and progressing through equipment installation, piping erection and so on
- The split into functional/geographical divisions such as the various process units, off-sites and utilities

Therefore, project strategy is notionally deciding how to cut up the cubic 'cake', the resulting 'pieces' representing the responsibilities of the owner, the EPCM contractor, the suppliers and construction contractors, with the way the pieces fit together being the interfaces in terms of stage, discipline and functional/geographical divisions. These 'pieces' representing work lots, become, as it were, the backbone of the project and are the basis on which purchase orders are placed and construction contracts are awarded (as well as providing the framework for planning and cost control).

Interfaces are potentially where mistakes occur, so they should not be excessive in number and should be placed logically in terms of physical works and documentation. Bitter experience shows that indiscriminate division of a project into a multiplicity of contracts with badly defined boundaries is detrimental to quality *inter alia* because the EPCM contractor's supervision is diluted over too large a number of suppliers and construction contractors with their associated interfaces.

1.5 Contractual environment

We can further examine the owner/EPCM contractor/ construction contractors model chosen in Section 1.3, with two alternatives illustrated in Figure 1.3. Option 1 shows a project where the EPCM contractor is responsible entirely and solely to the owner of the project, based on a lump-sum price.

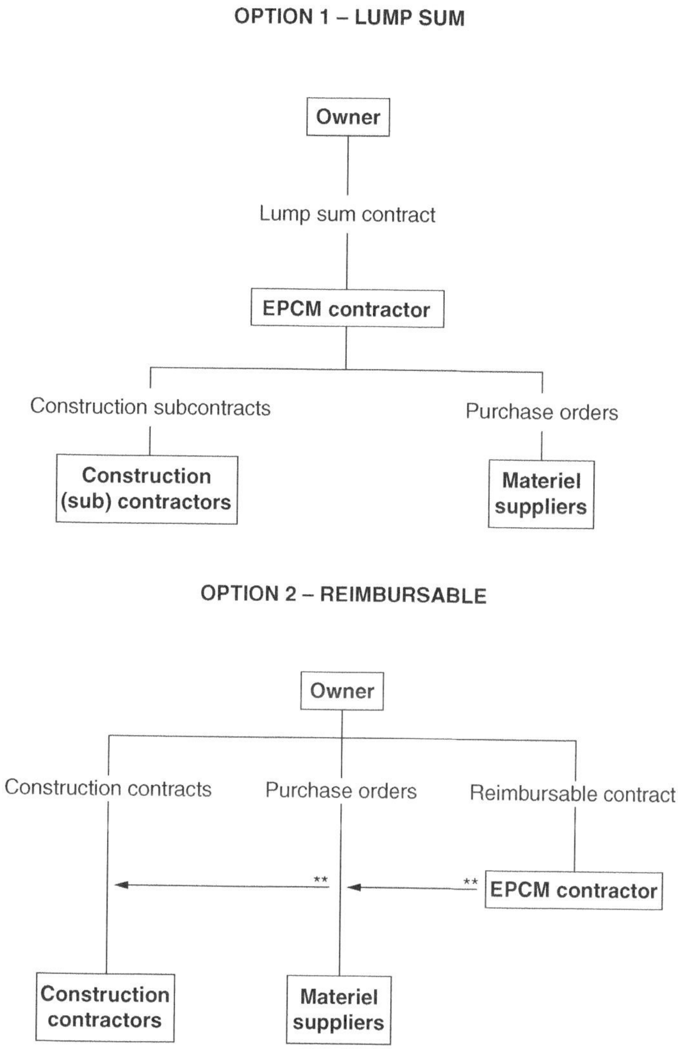

** Supervision of construction contractors and materiel suppliers by the EPCM contractor 'for and on behalf of the owner'.

Figure 1.3 Contractual configurations.

The EPCM contractor in turn places contracts directly with construction contractors, who in effect become subcontractors. Similarly purchase orders to suppliers for the provision of materiel are placed directly by the EPCM contractor. This arrangement, while it possesses apparent simplicity, requires the plant design to be defined in considerable detail before the appointment of the EPCM contractor. Option 2 presents an alternative multiple contract arrangement with the construction contractors and suppliers being contractually linked to the owner, but under the supervision of the EPCM contractor, acting for and on behalf of the owner. The EPCM contractor's contract with the owner is on a reimbursable basis. This arrangement enables the EPCM contractor to be appointed at an earlier stage in the project and gives greater flexibility throughout the project to meet the unexpected. A possible hybrid approach would be to start work on a reimbursable basis, then to change to a lump-sum contract when the design is sufficiently advanced.

Whatever the contractual roadmap of the project, the manner in which each contract is defined and administered can have a major influence, for better or for worse, on quality. Should an adversarial situation arise, the contractor will have his attention focussed on claims to the detriment of quality. The best way to minimize the risk of this happening is to ensure that the contract is fair and unambiguous. This is a statement of the obvious, but it is surprising how often this simple precept is ignored. You do not have to be a lawyer to understand that a contract will run smoothly and disputes will be avoided or at least attenuated if the following are clearly defined in writing and accepted by both parties from the outset:

- The precise nature and extent of the materiel and/or service in the scope of work to be provided
- The types of documents to be produced by both parties, and the procedure to exchange and process them
- How, where, when and by whom conformity to specified requirements is to be verified
- Requirements of authorities, if any
- Timing of the various stages, in particular time and place of delivery

- The price to be paid and terms of payment
- How variations in scope are to be processed

It is essential that agreements made during the negotiations, in particular concerning the resolution of differences between the call for bid documents and the offer, are formalized and made contractual as early as possible. Furthermore, the following situations must be avoided if a climate favourable to quality is to be generated and maintained:

- Imposition of unreasonably low prices
- Placing contracts late allowing no time for preparation
- Excessive modification in the course of the contract
- Not identifying and resolving quality problems as soon as they arise

To encourage the contractor to meet its obligations, nothing succeeds like attractive financial incentives. In practical terms, this means linking each important milestone, for example, the satisfactory completion of a significant element of the works, together with the associated inspections and tests, duly documented, to a substantial payment.

It has become commonplace in recent years to emphasize the 'togetherness' of the personnel of the various parties working on a construction site rather than the contractual relationships between them. This takes the form of holding 'team-building sessions' run by 'facilitators', leading to the drawing up of a 'partnership charter' expressing a common goal of finishing the plant under the best possible conditions, and so on and so forth. This is a laudable approach to improving communication between individuals coming from different horizons, who have to work closely together. It can generate motivation and mutual respect as well as create a project *esprit de corps*. However, it is naïve to imagine that such arrangements can lessen the necessity to build relationships on a bedrock of sound robust contracts.

2

ISO 9000

2.1 Introduction

ISO 9000 is a global term used to cover a whole family of standards relating to quality management systems, which are published by the International Organization for Standardization, but which also appear in other formats. The reader is recommended, in particular, to familiarize himself or herself with the two most important of these standards, namely:

- ISO 9000:2005 Quality Management Systems – Fundamentals and Vocabulary
- ISO 9001:2000 Quality Management Systems – Requirements

Other standards in the ISO 9000 series treat certain areas in more detail.

Why do we need ISO 9000? It gives a framework and a vocabulary widely recognized throughout the world for those creating and operating quality management systems. In fact, on most projects, for potential contractors to be entered on a list of bidders, they must be able to show that they have in place a quality management system complying with ISO 9001:2000. The availability of these standards, or their equivalents, with identical content in a number of different languages, can facilitate quality management on international projects.

2.2 Quality assurance

What are the benefits of applying a quality management system (QMS) as opposed to a traditional more informal approach simply relying on normal supervision? The application of an effective QMS, by its definition of procedures, its recording of inspection and test results, and its mechanisms for detecting and correcting anomalies, apart from its direct benefits, will give the necessary **assurance** that **quality**:

- From the very beginning and throughout the construction phase will be attained
- Upon completion has in fact been attained

Thus, all actions organized in a structured, systematic manner with this objective in mind can be described as **quality assurance (QA)**.

Effective QA enables the EPCM contractor to master quality issues, 'getting it right first time' with the resulting economies of time, money and effort. Externally, QA exercised by the EPCM contractor helps give confidence to the owner in the capacity of the former to deliver the plant to specified requirements. The same reasoning can be applied concerning suppliers and construction contractors. Their own QA measures produce internal benefits and, in addition, contribute to convincing the EPCM contractor and the owner as to their capabilities and the conformity of materiel supplied and works carried out.

2.3 Difficulties met

Unfortunately, the application of quality assurance (QA) in a construction context, and in particular the application of ISO 9001:2000, is often misunderstood or misapplied so that it does not realise its full potential. Let us examine these difficulties and see how they can be overcome.

In the first place, it must be understood that ISO 9001:2000 is a generic standard which can be applied to practically any institutional, commercial or industrial situation, for example, a school, a car hire agency, a washing machine factory or a construction site. So a considerable amount of effort has to be expended to translate the principles proposed therein into a format and expression relevant to a given context. In the case of this handbook, it has been written in the spirit of ISO 9001:2000, but expressed in terms which make sense on a process plant construction site.

Another problem which can arise is when a quality management system (QMS) proposed is too complicated for the activity or situation in question, with the texts of documentation couched in pompous terms, which practical men on construction sites have difficulty in taking seriously. To quote the reaction of one such person faced with this situation, 'QA is an expensive waste of time, generating unnecessary paperwork and stopping people getting on with the real work'. This state of affairs can be alleviated by top management exercising special care to recruit to posts of responsibility in the quality domain competent individuals familiar with the industry. Too often these positions are considered as token posts to be filled on a stop-gap basis.

Even though over the years effective formal QMSs have become more and more visible and seen to be relevant on site, there is still at all levels of the construction community, from operatives up to members of top management, a minority who are indifferent, sceptical or even hostile as to their relevance and utility. In answer to those who doubt, in the next section we set out from first principles the rationale for QMSs.

2.4 The rationale for formal quality management systems

Let us start by listing the essential conditions that must be met if quality of the finished plant is to be attained. These are:

- Responsibilities for attaining quality must be defined.
- Materiel arriving on site must be properly registered, stored, maintained and issued.

- The right documents must be available to the persons doing the work.
- A procedure needs to be in place to process site generated queries.
- Personnel shall be of appropriate levels of formation, skill and experience.
- Critical processes shall be the subject of predefined and validated procedures.
- Inspection, test and measuring equipment shall be regularly checked and calibrated.
- An appropriate level of traceability must be defined and implemented.
- The necessary physical and organizational means to successfully achieve the works shall be provided.
- A programme of inspection and tests shall be defined and implemented, with records kept of results obtained.

But surely meeting these requirements is simply good management, which any organization worth its salt would have being doing anyway? Yes, this is true and in the past this could have been handled in a relatively informal manner. However, today, with more and more complex projects, carried out to tight planning schedules and set within a strict regulatory framework, there has to be a structured approach. This means in the first place, creating and applying relevant written procedures. Let us now group together the procedures into what we can call a **quality management system** (QMS). However, the system will not come into being and work effectively on its own. It needs complete, wholehearted and visible support of **top management**. The latter will appoint a senior manager to be the guardian, champion and 'apostle' of the system 'to sell it' to the people who are going to have to make it work; let us call him or her the **quality assurance manager**. To sustain the system, a set of further procedures must be added to monitor its functioning, detect any anomalies and instigate corrective action, as well as to ensure its continuing improvement. At corporate level the complete definition of the QMS, together with a supporting statement from top management is presented in a **quality manual**. For a given project the manner in which the QMS will be applied to that project will be set out in a **quality plan**.

Using such an approach means that:

- We do not have to reinvent the wheel every time we open a new site.
- The site team can become fully operational very quickly.
- New staff can be rapidly introduced to project requirements.
- Nothing essential is forgotten.
- The existence of the system adds to the credibility of the organization and hence offers strategic advantage.
- A structure exists which can be continually reviewed and improved.

The above advantages are particular relevant to the construction industry which has to function in a difficult environment, for:

- Every project is to a greater or lesser extent a prototype.
- The workforce is nomadic, sometimes with little continuity in team composition from one project to the next.
- In some parts of the world, there is a limited industrial infrastructure, a shortage of skilled personnel, cumbersome administrative requirements and corrupt officials.

2.5 Certification of a quality management system

An organization may call on an accredited third party to assess its quality management system (QMS) with a view to its certification of compliance with the requirements of ISO 9001:2000. This in fact is a commercial necessity, for certification has become a prerequisite for entry on a list of bidders. However, in launching the certification process, it is important that the organization be motivated by a conviction of the intrinsic value of certification, rather than just one of commercial necessity. Existing resources may be available in-house to analyze requirements and then create or adapt the necessary procedures constituting the QMS. Alternatively, an organization may engage an outside consultant, in which case the selection of a suitable individual is primordial. The latter should be a practical down-to-earth person who, in addition to his or her in-depth appreciation of the ISO 9000 standards, is familiar with the construction industry and credible to site personnel.

2.6 Quality/occupational health and safety/ environment

Parallels can be drawn between systems for managing quality, occupational health and safety and environmental protection issues, which can be respectively structured to meet the exigencies of ISO 9001:2000, BS-OHSAS 18001:2007 and ISO 14001:2004. Each needs to be formalized in a site specific plan, to be implemented and operated on an ongoing basis and, above all, needs to receive the unreserved and visible support of top management. Moreover, each of these three management systems lends itself to the same monitoring, control and corrective action mechanisms implemented through the procedure for processing nonconformities, by auditing and through periodic management reviews. ISO 9001:2000 evokes the desirability for a quality management system to be compatible with systems covering other domains including occupational health and safety and environmental protection. Some situations on site may involve two or even all three areas at the same time. For these reasons, there is a tendency to rationalize and integrate the format, documentation and application of systems for managing these three domains.

3

On-Site Responsibilities and Interfaces

3.1 Introduction

Having already determined the project strategy as proposed in Section 1.4, which defines how the job is to be cut up and responsibilities for the various parts allocated, it is essential that at as early a stage as possible, the owner and the EPCM contractor formally review the project from start to finish in order to define relationships and interfaces. This joint exercise should include a review of the construction phase. Figure 3.1 outlines, in the form of a system, exchanges between the principal entities on site, that is, the owner, the EPCM contractor and the construction contractors. Also shown are the links to their respective head offices and with the utility companies, the insurer and the authorities, which together constitute the environment in which the site system is set. As the future arrangements take shape the discussion can focus onto a more detailed definition of 'who does what' on site in the form of a matrix, an example of which is shown in Table 3.1. This anticipates a more detailed treatment presented in subsequent chapters and sections.

This is a good place to remind ourselves that responsibility for effective quality management of construction is not just the concern of the EPCM site quality assurance manager and his or her staff, but implies the full and active cooperation

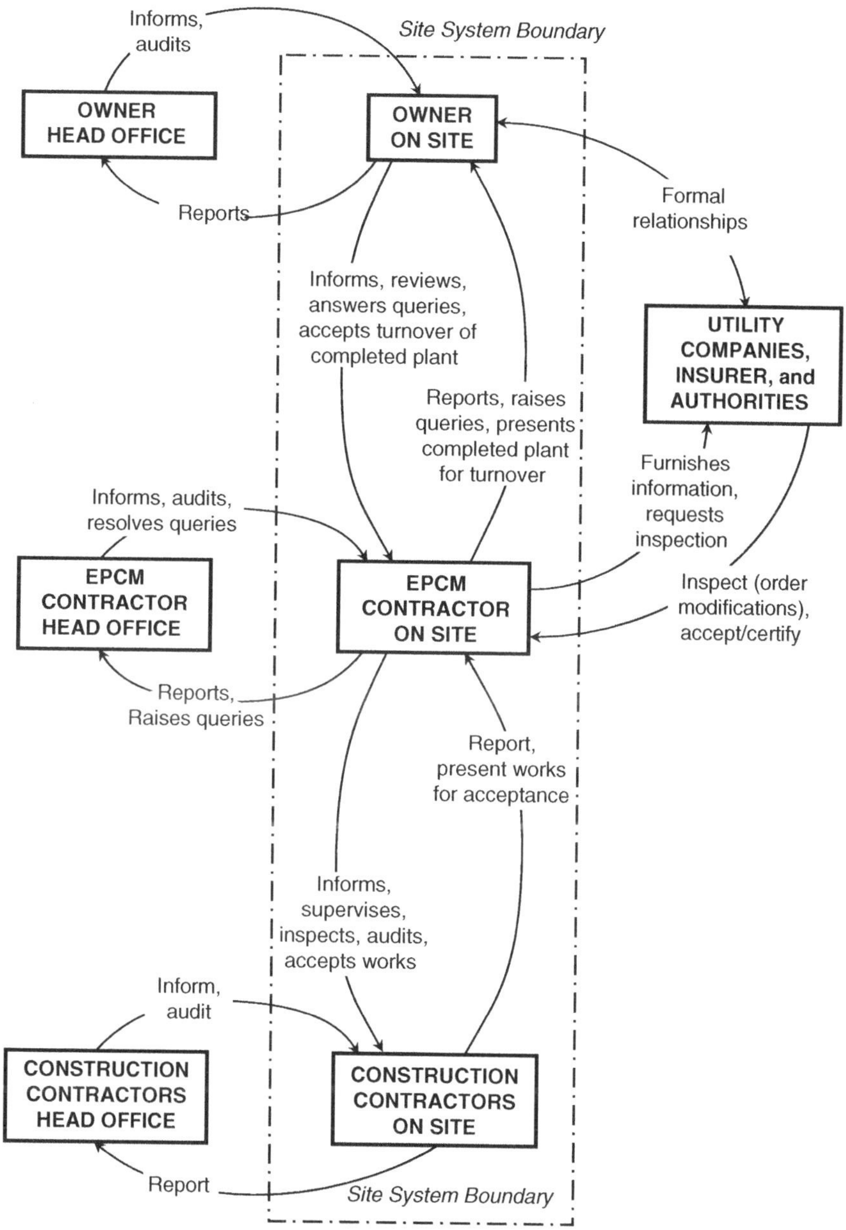

Figure 3.1 Site relationships – as a system.

Table 3.1 Site quality responsibility matrix in two parts (1/2 and 2/2).

EPCM contractor Name/Logo	SITE QUALITY RESPONSIBILITIES MATRIX Updated: ...../...../.....				Page: .....1/2...	Owner: .... Project no. .... Project:	
Activity/subject	**Construction contractor**	EPCM Contractor Department				**Owner**	**Reference procedure**
		Quality assurance	**Engineering***	**Materiel control**	**Construction**		
Site quality plan and inspection and test plan	Prepares	Reviews and approves	Reviews at QA request		Reviews at QA request		
Procedures and method statements	Prepares	Audits	Reviews and approves		Reviews at Engineering request		
Drawings 'approved for construction'	Receives and applies	Audits	Receives, registers and issues		Receives and supervises		
Materiel	Incorporates into the works and maintains	Audits		Takes delivery, checks, stores, maintains and issues	Maintains		
Contractor's workscope	Executes	Audits			Supervises		
Inspection, measuring and test equipment	Calibrates and uses	Audits			Checks		
Inspection and test records	Prepares and signs	Audits		Assists as required	Witnesses, checks and countersigns	May witness key tests	
Build-up of construction quality file	Files original, copy to EPCM contractor	Receives copy and audits			Supervises		
Engineering queries	May raise	May raise, audits	May raise, resolves		May raise	May raise	
Utilities, insurers and authorities		Audits	Technical Liaison		Assists for site visits	Formal relationship	
'As-built' drawings	Marks-up drawings	Audits	Receives, raises issue, manages		Supervises		

*Includes document control. Distribution:

Table 3.1 Continued

EPCM contractor Name/Logo	SITE QUALITY RESPONSIBILITIES MATRIX Updated: ...../...../.....				Page: .....2/2...	Owner: .... Project no. .... Project:	
Activity/subject	Construction contractor	EPCM Contractor Department				Owner	Reference procedure
		Quality assurance	Engineering*	Materiel control	Construction		
Nonconformities	May raise	May raise, registers and manages	May raise, may decide resolution		May raise	May raise	
Audits							
By the owner		Audited	Audited	Audited	Audited	Audits	
By the EPCM contractor head office		Audited	Audited	Audited	Audited		
By the EPCM contractor	Audited	Audits					
By the construction contractor head office	Audited						
By an accredited third party for ISO 9001certification of EPCM contractor	Audited	Audited	Audited	Audited	Audited		
Spare parts				Receives, stores, maintains and issues		Receives for start-up	
Construction completion by process unit or system by system	Presents as complete	Checks quality records	Checks "as-builts"		Checks works for completeness		
	Actions list of outstanding items				Makes list of outstanding items		
Turnover to owner by process unit or system		Presents	Presents		Presents	Accepts	

*Includes document control.

at all levels of all parties present. In the first place, the top management of the owner, the EPCM contractor and the construction contractors must want, and be seen to want, to make the quality management system (QMS) work as part of their company culture and strategy. They in turn must motivate and require their senior management on site to take the necessary steps to create and sustain the QMS, and then to encourage and motivate all levels of their staff including foremen and skilled tradesmen to collaborate fully in attaining quality objectives. On construction sites it is common practice to publish news sheets and create awards for achievements in matters of progress and safety. Doing the same thing for quality issues would help to sustain interest and motivation.

3.2 The owner

The owner is represented on site so as to be able to follow construction from day-to-day and to provide guidance to and answer queries raised by the other parties. The culmination of the construction phase is the turnover of the plant to the owner. This implies that when the time comes, the owner will be capable of accepting it. In practical terms this means that the owner will be building up the future team which will run the plant, in particular the engineering, production, maintenance and safety departments, to become progressively operational as construction completion and turnover are approached. The presence of key owner personnel on site during construction is an excellent means for them to become familiar with the plant in preparation for their future responsibilities. This is especially true during final inspection and testing of complete systems.

A close working relationship between the representatives of the owner and of the EPCM contractor on site is essential for the smooth working of the project. However, requests on the part of owner's personnel for alterations and additions must be formalized; otherwise they can get out of hand. The EPCM contractor must evaluate their technical impact as well as the impact on cost and delivery, and then the owner can decide whether or not to implement them. A vehicle for achieving this could be based on the site engineering query procedure proposed in Section 6.8, possibly with some adaptation.

3.3 The EPCM contractor

The EPCM contractor fulfils a key role in that it is the coordinator of the activities of all parties present and the catalyst to ensure the smooth running of the site. In particular, it selects the construction contractors and closely supervises their work. It is responsible for the issue of materiel and engineering documents to the construction contractors and ensures that the latter establish and continuously update quality records. The EPCM contractor identifies anomalies and launches the corresponding corrective action through a nonconformity management system, by auditing and by regular management reviews. It also liaises with other participants, in particular suppliers' representatives, special-service providers, utility companies, the insurer and the authorities, and is the main interlocutor on site for the owner. The ongoing objective of the EPCM contractor is to be able to turnover to the owner a complete plant to specified requirements (and of course on time and within budget).

Chapter 4 presents a more detailed discussion concerning the EPCM contractor on site.

3.4 Construction contractors

Construction contractors are chosen for their competence in the various disciplines. They bring to the site the necessary human and physical resources together with the structure, organization and procedures necessary to meet their contractual obligations. With the prior permission of the EPCM contractor, a construction contractor may subcontract part of its workscope, but in that case the construction contractor is entirely responsible for the quality of work of the subcontractor. The construction contractors receive engineering documents and materiel from the EPCM contractor, which they use to carry out the works. When required under the contract they may do some detailed design and procure materiel. They will have to submit a quality plan to the EPCM contractor for approval before putting it into operation. They shall perform the inspections and tests as per their quality plan, calling the EPCM contractor to witness when required to do so. Quality records shall be filed progressively to constitute a quality dossier. The ultimate aim of each construction contractor shall be to present the completed works for acceptance by the EPCM contractor, to be eventually turned over to the owner as a component part of the whole plant.

Chapter 5 presents a more detailed discussion concerning the construction contractors on site.

3.5 Suppliers' representatives

The more complex items of equipment, in particular complete packaged units, often require visits by supplier's representatives (SRs) to supervise the critical phases of erection, testing, commissioning and start-up. Their presence may even be continuous rather than on a visit basis. The purchase order will normally define the points at which visits are to be made and the costs and payment arrangements of such visits. Visits may be related to the question of guarantees, so it is important that they are organized on a sound basis. These services can be expensive, so the EPCM contractor needs to be firmly in control. This means:

- Designating an appropriate member of its staff as the interlocutor for each SR
- Making the SR report on arrival at site to define his or her scope of work and requirements in terms of assistance, materiel and services
- Requiring the SR to produce a written report before leaving site, this being a precondition for ultimate payment

This procedure can be implemented using the two-page form proposed in Figure 3.2.

EPCM contractor Name/Logo	**SUPPLIER SITE VISIT REPORT**	Report No. .... Page : 1/2

Owner: .... Project no. .... Project:	Supplier: .... P.O. no. .... P.O. : ....

This first page is to be jointly completed by supplier/EPCM contractor/owner at beginning of visit before the start of any work.

Supplier representative, ..., arrived on site at .... hours on ....
Expected duration of visit: .... days.

A. Object of visit

(Define which items of equipment are concerned)

B. Services to be provided by supplier's representative

C. Services/Material/Assistance to be provided to supplier's representative

(Utilities/scaffolding/manpower/etc.)

D. Permit to work required?

Yes/No*, Permit No.* ....

E. Cost of visit

Included in P.O./Is an extra*

F. Other remarks

*Delete as necessary.

For supplier	**For EPCM contractor**	**For owner**
Name:..........................	Name:..........................	Name:..........................
Signature:	Signature:	Signature: ....
Date:	Date:	Date: ....

Figure 3.2 Supplier site visit report in two parts (1/2 and 2/2).

EPCM contractor Name/Logo	**SUPPLIER SITE VISIT REPORT**	Report No. .... Page : 2/2

Owner: .... Project no. .... Project:	Supplier: .... P.O. no. .... P.O. : ....

G. Record of work carried out

To be completed by the supplier's representative before leaving site and to be countersigned by the EPCM contactor/owner. The following headings are to be used:

G1. Summary of actions taken	G4. Parts changed
G2. Details of modifications made	G5. Further action required
G3. Details of inspections and tests	G6. Conclusions/recommendations

Continue on extra sheet if required. Attach inspection and test records. Number of pages attached ....

	Report received, contents duly noted.	
For supplier	**For EPCM contractor**	**For owner**
Name:......................... Signature: .... Date:	Name:......................... Signature: .... Date:	Name:......................... Signature: .... Date: ....

Distribution:

Figure 3.2 (Continued)

3.6 Special-service providers

These organizations provide services such as:

- Surveying
- Site and laboratory testing associated with earthworks and concrete
- Destructive testing and nondestructive examination associated with welding
- The interpretation of results of such tests

Depending on project structure, these services may be provided within a construction contract or alternatively by special-service providers (SSPs) independent of the construction contractors. What is important is their availability, that is, to be on call to carry out a survey or test and to produce the report without delay. At remote sites, which may be a long way from suitable laboratories, it may be advantageous for the project to provide accommodation for the SSPs to be based on site.

3.7 Utility companies

Technical details concerning the supply of electricity, gas, water, telecommunications facilities and, where applicable steam, will normally have been settled before the site is open by collaboration between the utility companies, the owner and the EPCM contractor's engineering department, formal contractual arrangements being the responsibility of the owner. Besides supply arrangements to the future plant, temporary utilities will be needed during construction. It is important that site visits of representatives of the utility companies be arranged in a timely manner so that any potential problem areas are identified early and not left to be discovered at the last minute, possibly delaying commissioning and start up. Utility companies are interested in their interfaces with the plant, that is, incoming switchgear or main isolating valves and the arrangements for metering. In addition, they will want to be sure that the installations within the plant will be compatible with their supply networks, concerning protective relay discrimination for example.

3.8 The insurer

The owner's insurer's premiums will be determined according to its assessment of risk, so this organization will have defined requirements concerning certain aspects of the plant to be incorporated into the design. The insurer may well also want to be involved at the construction phase. So it will be necessary to define which documentation will be needed for review and which tests will the insurer wish to witness. An early visit to site by a representative of the insurer is necessary so that all parties understand what is required.

3.9 Authorities

It is essential that the plant meets the legal and regulatory requirements in the country (or countries) of manufacture of materiel and the country in which the future plant will be situated. The most effective way to ensure that nothing is missed is for the EPCM contractor to appoint at the beginning of the project an authorities engineer, who shall identify the exigencies of each authority ensuring that these requirements are incorporated into the plant design. From then on he or she shall, in collaboration with the owner, coordinate the authorities':

- Design approvals
- Visits to suppliers
- Site visits

During construction it might be preferable for the authorities engineer to be based on site. When the site is in a country with which the EPCM contractor is not familiar, it is essential to have an input of local knowledge. It is common practice also for the owner to appoint a third-party organization to review plant design, followed by surveillance during subsequent operations, including construction and the related inspection and test programme, to ensure compliance with legal and regulatory requirements. The status of such a third party would depend on the project in question and the country in which it is situated. In this handbook any such third party is deemed to be covered by the generic term 'authorities'. We can extend the notion of 'authorities' further to include not only central government departments or their chosen representatives, but also the local authority, police and fire brigade, which may well have jurisdiction in one way or another over the site. The watchword is to establish a good working relationship with these entities because you may need their goodwill.

There may well be considerable overlap between the responsibilities of the utility companies, the insurer and the authorities. For example, the electricity supply company may be the *de facto* or even the *de jure* authority for the plant electrical installation. In some countries the enforcement of regulations concerning boilers and lifting devices may in effect be left to the insurer. For this reason it may be advantageous for the EPCM contractor's authorities engineer also to be responsible for liaison with the insurer and the utility companies.

3.10 The architect

All projects, even those designed for outdoor operation, include buildings such as offices, stores, workshops, a canteen, an infirmary, cloakrooms, toilets, control rooms, electric substations and gatehouses. A rational conception of these buildings contributes to the future effective operation of the plant and to good company–workforce relations. They are generally subject to mandatory conformity to building codes and in addition are the shop window of the plant to the outside world. For these reasons it is highly desirable, often obligatory, to employ a qualified architect; he or she could be from within the EPCM contractor's own organization or could be an independent professional. Apart from the architect's role in conception, detailed design and choice of materials, his or her presence on site can contribute to the satisfactory achievement of the building programme.

3.11 Plant extensions and modifications

When a project is an extension or modification of an existing process plant, rather than the creation of a new one, the ground rules have to be adapted accordingly. In the first place the owner is omnipresent and the EPCM contractor's staff on site must work in close harmony with the various operational departments of the plant. For example, new construction work would probably have to be executed for reasons of safety within the existing permit to work procedure. This is even more the case when the works are in effect embedded in the existing plant rather than situated in a distinctive adjacent zone. Engineering must produce drawings clearly identifying the interfaces between the existing installations and the future ones, and this in all disciplines. Then there must be a clear definition of modifications to be made to the existing plant and the manner in which new and existing elements are to be tied-in. As far as possible any duplicate items of equipment should be procured from the original suppliers. Also ideally the same EPCM contractor and construction contractors should be used, especially if senior staff who participated in the initial project could be made available, though this might not be attainable in practice. Various operations, in particular tie-ins, have to be accomplished according to a strict timetable within narrow opportunity windows involving partial or complete plant shutdown.

Thus the extension/modification of an existing plant calls for meticulous attention to detail in pre-planning, in execution and in completion. Written procedures for all phases must be prepared well in advance and accepted by those concerned. The necessary staff, in particular in the engineering disciplines, must be available on site for monitoring and control. The respective responsibilities of EPCM contractor and owner have to be defined especially for construction completion, turnover and commissioning. Finally the project organization must include provision, which means the necessary funds and staff being made available, to integrate new and existing documentation into a coherent whole to facilitate operation and maintenance of the future extended/modified plant.

3.12 Small construction sites

The smaller construction site will generally be given a simplified management structure; in particular there may not be present on site a distinctive person or team to create and manage on a day-to-day basis a formal quality management system. However, whether these persons are present or not, the quality issues are still very much in evidence, because the objective must always be to complete construction and handover the works in conformity with specified requirements to facilitate subsequent commissioning and start-up. The other idea to bear in mind is that a quality management system is not, or should not be, a heavy unyielding mass weighing down a project. It is rather of the very essence of the project directed towards simplifying rather than complicating the aim of advancing towards completion of construction to the satisfaction of all parties. This notion is developed in more detail in Section 2.4. So the question is not one of supposing that on a small site it is not possible to impose a quality management system, but rather one of how quality objectives can be achieved with the reduced means available.

This must be discussed, decided and documented before the site is opened. Some or all of the following approaches can be used to simplify structures and procedures:

- Integrate owner and EPCM contractor site management.
- Reduce the number of construction contractors, and combine related disciplines into one contract, mechanical and piping, or electrical and instrumentation and control, for example.
- Include in the construction contracts procurement and warehousing of material, pipes and fittings, or electric cables, for example.

- Where possible maximize the use of prefabrication with stringent testing and inspection under factory conditions before delivery, thus reducing site works and the corresponding quality concerns.
- Use known and tested construction contractors who will attain quality goals with minimal supervision on the part of the EPCM contractor.
- Completely define and document the inspection and test programme with the corresponding set of record forms to be available the day the site opens, and make sure that the construction supervisors implement the programme.
- Assign, train and motivate someone to run on a part time basis the nonconformity management system.
- At the time of opening the site hold a workshop for all concerned individuals run by a competent quality professional and with the presence of top management, to explain what is required and to motivate everyone.
- Arrange for this quality professional to return to the site at regular intervals to audit the proceedings, proposing corrective action for any shortcomings discovered.

Most important of all is that senior management of all parties on site are motivated, believe in the validity and relevance of the simplified quality management system put in place, want to make it work and are able to transmit their commitment to their staff at all levels.

4

The EPCM Contractor

4.1 Introduction

This chapter examines how the EPCM contractor can organize its task on site, first of all in creating its quality plan and then in determining the composition of the site team. The attributes of key members of staff, whose actions and attitudes are crucial for attaining quality objectives, are then evoked, followed by a study of the division of responsibilities for quality control.

4.2 The corporate quality manual

It is probably true to say that most EPCM contractors possess a corporate quality manual, which, as the name implies, covers all of the organization's activities, structures and procedures ranging from those which are continuous and those which are shorter-termed ones created or adapted for use for the duration of a project. This document represents the organization's accumulated collective know-how on quality management. Supporting documents are either included or reference is made to them.

4.3 Creating the site quality plan

In an ideal situation the corporate quality manual (CQM), evoked in Section 4.2 above, should include a complete set of site-related model procedures and typical organizational schemata so that a project specific site quality plan (SQP) can be created simply by adapting this information to the specific needs of the construction site in question. Unfortunately, all too often in the CQM, the construction site element is lacking in content and format. So what can we say to the individual who has to create a SQP at short notice without the backup of adequate recorded accessible corporate know-how? He or she could start out by reading ISO 9000:2005 and ISO 9001:2000 for background information (see Chapter 2). The terms and definitions given in Section 3 of ISO 9000:2005 and the relationship diagrams presented in Annex A are particularly relevant. He or she could then, in consultation with site management, work through this handbook adapting the principles and model procedures/forms set out herein to the particular needs of the site being considered. This approach can be summarized thus:

- Identify the interested parties (Section 1.3).
- Identify how the retained project strategy has divided up the responsibility for the various phases of the project, in particular for construction (Section 1.4 and Figures 1.1–1.3).
- Define the role of each interested party on site and how the parties present relate to each other and with entities exterior to the site (Chapter 3 with Figures 3.1 and 3.2, and Table 3.1).
- Define the EPCM contractor's site organization (Chapter 4 and Figure 4.1) and in particular responsibilities for quality control (Figure 4.2).
- Define what is required of the construction contractors, including the keeping of quality records (Chapter 5).
- Define the role of the EPCM contractor's engineering department on site, including document control, and the site's interface with the project information technology infrastructure (Chapter 6).

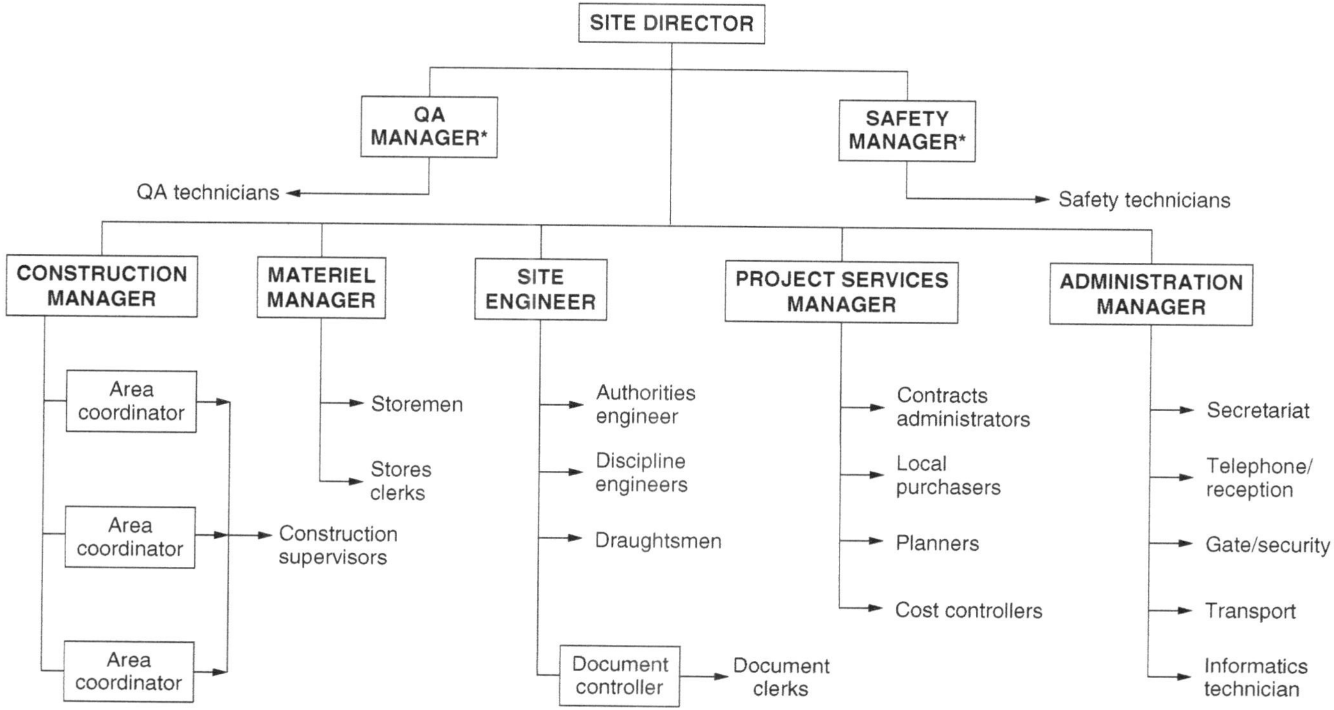

*One or other may also be responsible for environmental issues.

Figure 4.1 EPCM contractor – typical site organization chart.

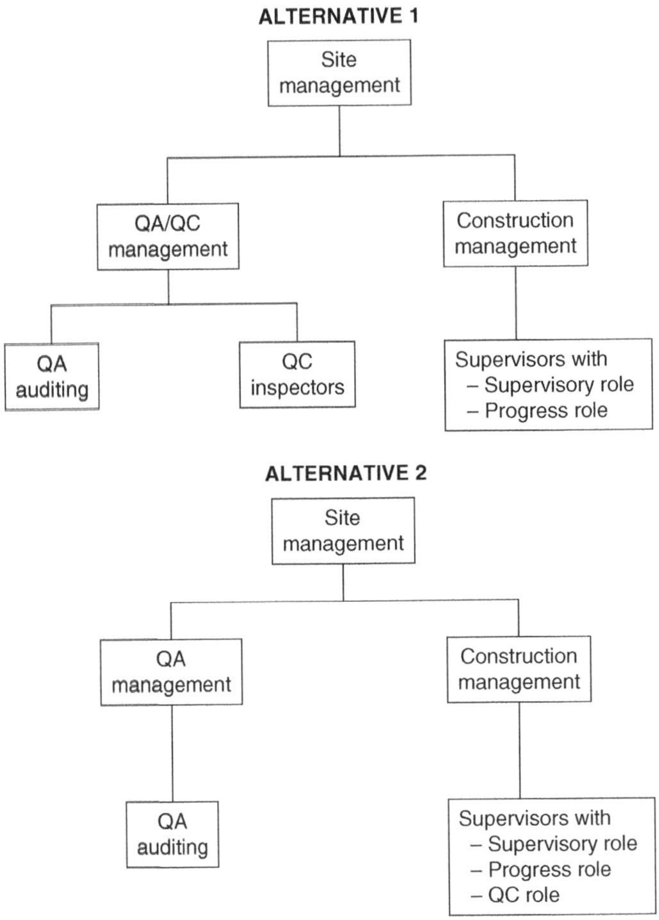

Figure 4.2 EPCM contractor – quality control responsibilities.

- Establish a materiel management procedure (Chapter 7).
- Define construction completion and turnover, and possible technical assistance to the owner beyond turnover (Chapter 11).

The above constitutes the main working elements of the EPCM contractor's site quality management system, but how do we monitor, control and if necessary adjust and correct the system? There are three correction and feedback tools, namely:

- A nonconformity management procedure (Chapter 8)
- A quality auditing procedure (Chapter 9)
- A management review procedure (Chapter 10)

As construction draws to its completion and the works are soon to be turned over to the owner, it is highly desirable to carry out a systematic review of the whole site process in order to be able to identify and record lessons to be learnt and feed them back into the corporate management system in the form of a construction completion report as part of a policy of continuous improvement (Chapter 10).

The documents resulting from the above exercise can be grouped together to constitute the SQP, to be validated by top management before being issued and applied. The SQP has to be compatible with procedures concerning administration, cost control, planning, occupational health and safety, environmental protection, site procurement and so on. In fact, some EPCM contractors create one single project site execution manual, embracing all facets of site management including quality. This has the advantage of presenting quality as a mainstream concern rather than as a bolt-on extra. It goes without saying that all of the above cited procedures need to be created in liaison with and taking into account the remarks and views of those who will have to apply them.

4.4 The organization chart

Figure 4.1 shows a typical site organization chart for an EPCM contractor. The number of persons present depends on the size, type and structure of the project. Job titles vary from one organization to the next, but whatever the titles, all of the functions shown must be covered in one way or another. The following are brief comments on certain posts shown in the organization chart which have a significant impact on quality issues:

Site director. This person must give his wholehearted and visible support to the quality management system (see Section 4.5 for a more detailed discussion).

Quality assurance manager. This is a key position with responsibility for overseeing the functioning of the whole site quality management system (see Section 4.6).

Construction manager. This post is totally devoted to the supervision of the works carried out by the construction contractors (see Section 4.7).

Site engineer. This is the site representative of the EPCM contractor's head office engineering department; responsibilities can include document control (see Chapter 6).

Materiel manager. Responsibilities of this post are essentially receiving, storing and issuing of materiel on site (see Chapter 7).

4.5 The site director

The role of the site director is to manage the site in liaison with all interested parties such that construction is completed on time, within budget and to specified requirements (quality). There is an increasing tendency to compress the time available for realization, in particular telescoping the construction phase, the buzzword for this being 'fast-track'. So the site director is going to be under a lot of pressure to meet the construction completion deadline. However, there is really no point in going through the motions of completing construction on time, or even ahead of time, as well as remaining apparently within budget, if along the way a series of grave, unresolved quality problems have accumulated. The latter will have to be corrected with the subsequent delays and costs, wiping out any previous supposed advantage, as well as generating concern on the part of the owner. For this reason it is vital for the site director to be of sufficient personal and professional calibre to resist pressure to deliver willy-nilly. He or she must, instead, give to quality its appropriate status in the cost–delivery–quality triangle. So the right person to fill this post will be someone, who apart from his or her other attributes, will from the very beginning give unreserved support to the creation and implementation of the site quality management system. The site director shall above all perceive the latter as a vital tool to achieve project objectives (including the mastery of delivery and costs) and not just a chore imposed by head office.

4.6 The site quality assurance manager

The site quality assurance manager (SQAM) is the person with over all responsibility for creating and implementing the site quality management system. He or she should be an engineer or senior technician possessing a sound general technical and professional education, followed by broad-based work experience. Strong verbal and written communication skills, as well as the capacity to grasp the significance of matters across the whole spectrum of activities and situations present on a process plant construction site, are a *sine qua non*. The SQAM must in addition be diplomatic and be perceived by others as credible.

4.7 Responsibilities for quality control

Quality control (QC) encompasses surveillance, inspection and testing. Before going any further, let us define these terms. **Surveillance** means monitoring the works on an ongoing basis to detect in a timely manner potential and existing problems. **Inspection** is the formal verification of works presented as being in conformity with specified requirements. **Testing** is the formal verification on the basis of an established procedure that the materiel or works presented possesses a defined characteristic. **QC** is the collective term covering all three.

As a general principle it is important that the maximum responsibility for quality be exercised as close as possible to those carry out the physical work. It is a mistake to suppose that quality can be 'inspected into' the works. Instead, the underlying basis for achieving quality is to deploy skilled motivated workers under the surveillance of competent supervisors. The fundamental role of the construction contractors is to do just that. The latter shall only present to the EPCM contractor for inspection and witnessing of test the works already deemed to be in conformity with specified requirements.

So this leads us to the question of how and by whom QC is managed within the EPCM contractor's organization. Figure 4.2 presents two alternative schemata:

- Alternative 1, with QC independent of construction and instead combined into an integrated QA–QC function.
- Alternative 2, with QC integrated into the construction supervision function.

Alternative 1 gives a powerful independent quality function which certainly has advantages. However, with this arrangement 'quality' can develop into an independent 'empire' in parallel with construction supervision. The resulting relationship can be conflictual.

Alternative 2 gives the EPCM contractor's construction supervisors global responsibility for the execution of the works, including supervision of the construction contractors' programmes of inspection and tests (all subject to audit by the QA function). This second configuration has the advantage of placing responsibility for QC closer to the day-to-day action, but still independent of the construction contractors. Generally the second configuration is preferable, but it can only work if the construction supervisors are prepared to assume responsibility for quality, including ensuring that proper formal records are kept of inspection and test results. For reasons specific to the culture of the industry this is not always easy to attain, but with the new generation of more educated, trained and motivated individuals arriving on the scene, it is perfectly feasible. This latter arrangement is the one assumed in this handbook.

CASE STUDY

Let us compare the very different cultures of two EPCM contractors, 'X' and 'Y', concerning the management of quality on their process plant construction sites.

In the X organization there is a corporate quality assurance manager who regularly audits both office and site activities. At the close down of each site, a comprehensive report is written covering all activities, including the various elements constituting the quality management system, evoking successes attained and problems experienced, with recommendations for future projects. This recorded experience is used to update the corporate construction quality manual. The latter includes a range of typical procedures and forms, providing the basis on which a project specific quality management system can be created in a timely manner. A quality assurance manager is appointed on each site, where quality issues are considered a central element in day-to-day management. X is certified to ISO 9001:2000; being re-audited to extend the certification does not present a major problem. Whilst X's quality management performance on its sites is not perfect, the fact that 'quality' is a major preoccupation of the company's top management means that 'quality management' in an integral part of every project and is accepted as such.

Y's top management prides itself in the ability and experience of its long serving engineers and technicians, but does not attach much importance to creating a bank of formal corporate procedures. Consequently every time a new process plant site is opened, the wheel has to be reinvented, but this is not considered to be a problem since the staff 'are competent and will thus succeed in doing whatever is necessary, so they don't need procedures'. Under commercial pressures Y has gone through the process of obtaining ISO 9001:2000 certification but without any real enthusiasm, so it is not surprising that when the audit for renewal of the certification is a long way off, the quality management structure tends to evaporate. In spite of this negative ambiance, individuals on one site managed to create and made to work in an effective manner major elements of a quality management system,

Continued

CASE STUDY (Continued)

such as a site quality plan, inspection and test plans, quality records, nonconformity management and audits, but when the project finished there was no corporate structure to record and pass on the useful experience gained for the benefit of a subsequent project. Y has a corporate quality assurance manager, but his role is not allowed to be much more than that of a facade to be presented to potential clients.

Lesson learnt

Here, there is only one fundamental lesson to be learnt upon which everything else depends and that is that if a quality management system is to become an integral part of a company's culture, top management must consider it essential and will in consequence provide the necessary resources and ongoing dynamic support.

5

Construction Contractors

5.1 Introduction

Construction contractors are the organizations which physically carry out the work on site. In fact, it is no exaggeration to say that the success of the construction phase of a project is very much bound up with the selection of competent contractors, engaging them under sound contractual conditions and creating an environment on site to enable them to fulfil their role to the satisfaction of all parties. With this in mind it is vital to define from the outset what is expected of a construction contractor concerning quality management.

Construction contracts can be placed on the basis of covering a single discipline, or by grouping within the one contract two or more closely associated disciplines indicated with asterisks as follows:

- Civil works*
- Building*
- Structural steelwork*
- Mechanical erection**
- Piping**
- Electricity***
- Instrumentation***

- Painting****
- Thermal insulation****

In addition, some contracts may be let on a multi-disciplinary basis, each being in effect a project within the project. Examples are complete buildings, water treatment plants and electric substations.

5.2 Selecting the construction contractors

What are we seeking in a potential construction contractor? As far as quality on site is concerned, we are looking for:

- A proven record of successful construction operations on sites comparable to those of the project in hand
- Established simple, coherent and realistic corporate procedures and working methods, appropriate to and easily adaptable to the specific needs of the project
- The above-mentioned procedures and working methods set within the context of a quality management system, which is a working reality, fully supported by top management and not just a set of glossy folders gathering dust on a bookshelf
- A pool of experienced individuals available to be deployed on the forthcoming site
- Physical resources, ample and appropriate for the needs of the project

How do we find out to what extent a potential construction contractor possesses these attributes? We can take all or some of the following actions:

- Visit the sites that have been completed by the potential contractor, better still visit a site or sites where it is actually engaged in construction
- Talk with clients past and present
- Visit corporate headquarters and review the company's procedures and working methods

- Review the company's quality management system in the light of certification to a recognized quality management standard, if necessary by means of a formal audit
- Interview key personnel presented to fill the posts on site in the event of a contract award

All the above investigations should be the subject of written reports. The ensemble of these documents can then be analysed to give an overview as to the aptitude of the contractor to be able to respond to the needs of the project. A preliminary assessment can be made at the time of drawing up the list of bidders and a deeper investigation made as part of the bid evaluation of say the two or three most likely candidates.

5.3 The initial site meeting with each construction contractor

As soon as the contract has been awarded and the construction contractor has mobilized its staff on site, an initial meeting must be held with the various departments of the EPCM contractor to establish lines of communication, clarify any anomalies and ensure that both parties put in hand the necessary arrangements for the smooth working of the contract. At this sort of meeting questions of quality tend to take a low priority and hence get squeezed in at the closing stages of the meeting after safety, administration, invoicing, and so on. It is preferable, instead, for the participants of both parties who are concerned with quality matters to hold a separate parallel meeting and then to report their conclusions to the main meeting. This enables quality questions to receive the attention that they merit. It is essential that minutes be written as the meeting advances and then at the end signed by both parties. This ensures that no doubt subsists as to what has been agreed, even though the formal typed minutes may be signed and distributed later. Against each action the person responsible for its implementation shall be named and a date defined for its completion.

A model agenda for the initial site meeting is proposed hereunder. It can be adapted to the needs of a specific contract:

- Responsibilities and interfaces on and off site
- Subcontractors
- Special-service providers
- Requirements of the contract and related requisitions, specifications and drawings
- Construction contractor's site quality plan*, contents and date for presentation
- Inspection and test plan*, specimen forms for inspection and test records*
- Weld procedure specifications* and procedure qualification records*
- Other procedures and method statements*
- Qualification of welders, nondestructive examination personnel and any other categories of personnel requiring formal qualification
- Traceability
- Calibration of inspection, measuring and test equipment
- Document control
- 'As-builts' drawings
- Engineering queries resolution
- Information technology interfaces
- Materiel control
- Management of nonconformities
- Auditing
- Construction quality file*
- Construction completion
- Any other matters

The asterisks identify documents for which timely presentation is essential to the smooth running of the contract. The construction contractor shall be encouraged to deliver these on time by linking them to substantial payments defined contractually.

Once the construction contractor is working on site, there will be regular progress meetings with the EPCM contractor. On the same occasion parallel 'quality meetings' can be held. Items can be selected from the above agenda to establish a standard agenda for such meetings. It is particularly important that at each of these regular meetings the questions of outstanding nonconformities and audit findings are discussed and expedited to close out.

5.4 Site quality plan

The construction contractor's site quality plan (SQP) is a contractual document to be prepared and presented to the EPCM constructor's site quality assurance manager as soon as possible after the arrival of the construction contractor on the site. It is a brief straightforward document that defines and explains how the contractor proposes to manage quality issues on site; that is to say, it is the roadmap of his quality management system. Chapter headings could be:

- Contract Workscope Breakdown
- Site Organization Chart
- Quality Assurance/Quality Control Functions
- Document Control
- Materiel Control
- Identification, Marking and Traceability
- Procedures/Method Statements
- Personnel Records
- Physical Resources
- Quality Records
- Nonconformity Management
- Quality Audits
- Inspection, Measuring and Test Equipment
- Other Relevant Details
- Validation

Referring to the four-paged Figure 5.1, there will be found a model SQP form where the above-listed chapter headings are set out with explanations and space to be filled in to complete the document. This form can be:

- Included with the bid documents to indicate the type of SQP expected in case of award of contract

<table>
<tr><td colspan="2">EPCM contractor

Name/Logo</td><td colspan="2"><u>CONSTRUCTION CONTRACTOR</u>

<u>SITE QUALITY PLAN</u></td><td>SQP no. ....
Rev: ....
Date: ....
Page: 1/4</td></tr>
<tr><td colspan="3">Owner: ....
Project no. ....
Project: ....</td><td colspan="2">Contractor: ....
Contractor no. ....
Contract: ....</td></tr>
<tr><td colspan="5"><u>A. CONTRACT WORKSCOPE BREAKDOWN</u></td></tr>
<tr><td><u>Contractual status</u></td><td><u>Name of organization</u></td><td><u>Location</u></td><td colspan="2"><u>Services provided/Workscope</u></td></tr>
<tr><td>Main contractor

For joint venture: Other partners

Sub contractors

Others</td><td></td><td></td><td colspan="2"></td></tr>
<tr><td colspan="5"><u>B. SITE ORGANIZATION CHART</u>
(Include interfaces with main contractor's head office)</td></tr>
<tr><td colspan="5"><u>Distribution:</u></td></tr>
</table>

Figure 5.1 Construction contractor – site quality plan in four parts (1/4, 2/4, 3/4 and 4/4).

EPCM contractor Name/Logo	**CONSTRUCTION CONTRACTOR** **SITE QUALITY PLAN**	SQP no. .... Rev: .... Date: .... Page: 2/4

C. QUALITY ASSURANCE/QUALITY CONTROL FUNCTIONS

(Who is responsible for QA/QC on site and in head office? Describe their functions)

D. DOCUMENT CONTROL

(Describe document control function on site with respect to documents received from the EPCM contractor and/or generated by the contractor itself. Refer to separate procedure if appropriate and attach a copy as an appendix)

E. MATERIEL CONTROL

(Describe the materiel control function on site with respect to materiel free issued by the EPCM contractor and also concerning materiel procured directly by the contractor. Refer to separate procedure if appropriate and attach copy as an appendix)

Space out sections as required

F. IDENTIFICATION, MARKING AND TRACEABILITY

(Define principles. Define traceability trails. Refer to separate procedure if appropriate and attach a copy as an appendix)

G. PROCEDURES/METHOD STATEMENTS

(Give a complete list of all relevant documents, including those for special processes such as welding, surface treatment, etc. Include a copy of each as an appendix)

DOC. No. REV	DOCUMENT TITLE

Figure 5.1 (Continued)

EPCM contractor Name/Logo	**CONSTRUCTION CONTRACTOR** **SITE QUALITY PLAN**	SQP no. .... Rev: .... Date: .... Page: 3/4

H. PERSONNEL RECORDS

(List classes of personnel record to be established and maintained, in particular for welders, NDE technicians, electricians, ...)

I. PHYSICAL RESOURCES

(List construction equipment, vehicles, office and IT facilities)

J. QUALITY RECORDS

(List inspection and test plans, inspection and test record forms for the various quality control operations to be carried out by the contractor. Include a copy of each document as an appendix)

Space out sections as required

K. NONCONFORMITY MANAGEMENT

(The EPCM contractor will provide a site wide system, but give details of any internal system)

L. QUALITY AUDITS

(The contractor on site will be audited by the EPCM contractor, but give details of any internal audit programme)

Figure 5.1 (Continued)

EPCM contractor Name/Logo	**CONSTRUCTION CONTRACTOR** **SITE QUALITY PLAN**	SQP no. .... Rev: .... Date: .... Page: 4/4

M. INSPECTION, MEASURING AND TEST EQUIPMENT

(List items and define regular calibration programme)

N. OTHER RELEVANT DETAILS

(Add further relevant details of contractor's site quality managemnet system)

Space out sections as required

O. VALIDATION

This document with attached appendices as listed below constitutes the site quality plan applicable to all works carried out by the construction contractor, ... on or around the site for the execution of the above defined contract. After approval by the EPCM contractor, it will be issued for immediate application.

For: ....

Name: ...., at ....

Signature: .... Date: ....

LIST OF APPENDICES

Figure 5.1 (Continued)

- Used as a questionnaire during a bid clarification meeting and form the basis of a report
- Used by the construction contractor as a guide for preparing the SQP to its own format
- Filled in directly on site by the construction contractor, the completed and signed document constituting its SQP

The last use has proved to be a convenient and rapid way to establish a SQP, and in that it has been created on site by the individuals concerned, it may be more readily acceptable to them than a document imposed by their head office. It is also relevant to the smaller construction contractor, whose site organization will be minimal and whose corporate culture tends to be more informal than that found in a larger organization (see Section 3.12).

5.5 Inspection and test plan

The purpose of an inspection and test plan (ITP) is to summarise and present in a user friendly tabular form the complete programme of inspections and tests for the whole contract scope or for a section thereof, based on information extracted from the relevant drawings, specifications, codes, standards and regulations. It is intended for day-to-day ready reference on site. The ITP can be produced contractually by the construction contractor, which has the advantage that it will in effect 'own' the document and benefit from having produced it. Alternatively, it could be produced by the EPCM contractor. This latter option would be appropriate in a fast-track project where there would be little time to prepare the document before putting it into application.

A model ITP form is presented in Figure 5.2. An explanation of the purpose of the first six columns is given hereunder:

- *Column 1* – Enables the inspection and test points to be numbered in chronological order
- *Column 2* – Definition of inspection or test point
- *Column 3* – Frequency, such as 'once per shift' or '10% of welds'

<table>
<tr><td colspan="2">EPCM contractor

Name/Logo</td><td colspan="4">INSPECTION AND TEST PLAN

ITP no. .... Updated: ...../...../..... Page: ....../......</td><td colspan="2">Contractor: ....
Contract no. ....
Contract: ....</td><td colspan="2">Owner: ....
Project no. ....
Project: ....</td></tr>
<tr><td rowspan="2">Point no.</td><td rowspan="2">Inspection/Test point</td><td rowspan="2">Frequency</td><td rowspan="2">Reference document</td><td rowspan="2">Acceptance criteria</td><td rowspan="2">Record format</td><td colspan="2">Hold point</td><td rowspan="2" colspan="2">Remarks</td></tr>
<tr><td>EPCM</td><td>Owner</td></tr>
<tr><td></td><td></td><td></td><td></td><td></td><td></td><td></td><td></td><td colspan="2"></td></tr>
<tr><td colspan="10">Distribution:</td></tr>
</table>

Figure 5.2 Inspection and test plan.

- *Column 4* – Document and clause number defining requirement
- *Column 5* – Acceptance criteria, such as '60 bars for 1 hour' or 'deflection <1/ 750 of span'
- *Column 6* – Defines the form or other document on which inspection and test results are to be recorded

Columns 7 and 8 allow witness and hold points to be identified by the various parties, the first column being for the EPCM contractor, the second for the owner, insurer or authority according to need. Letters can be added in each of these columns to indicate:

- W = 'Witness point' means that the party so indicated will be given notice of the inspection or test, which shall proceed at the appointed time, whether the party's representative is present or not.
- H = 'Hold point' means that the party so indicated must be present and the inspection or test shall not proceed without it.

The number of hold points should be kept within reasonable bounds. If there are too many, it will be difficult to honour them all. This can lead to delays and friction. In some contexts quality can be adequately verified by simply inspecting and/or testing the finished works, but more often additional key inspection/test points have to be fixed at intermediate stages as the installation or construction advances, it being physically impossible to carry them out later.

5.6 Procedures and method statements

All activities having a significant impact on the quality of the works, and where information given in the contractual drawings and specifications is insufficient, shall be the subject of a procedure or method statement to be submitted by the construction contractor to the EPCM contractor for review and acceptance before the start of the works. In the specific case of welding, these take the form of weld procedure specifications and the associated procedure qualification records. Other procedures and method statements may be necessary, for example, for operations such as casting of major concrete structures, heavy lifts, reservoir tank hydro tests and settlement checks. However, documents which merely paraphrase existing specifications are superfluous, lead to confusion and are thus to be actively discouraged. Each construction contractor shall set out in his or her SQP a list of procedures and method statements that it will be producing. In fact, for the more important ones, their creation, submission for approval and their eventual application should be contractual and the subject of a specific payment.

5.7 Inspection and test records

At appropriate points throughout the course of the works, the construction contractor must propose and organize the inspections and tests defined in the inspection and test plan (ITP) mentioned in Section 5.5, giving appropriate notice to the EPCM contractor and any other parties indicated on the ITP for the operation in question. Inspection or test results obtained as well as the signatures of parties present/reviewing the document must be entered on inspection and test record (ITR) documents. Their format and contents shall be agreed upon before the start of the works. Ideally a construction contractor will already possess in its corporate procedures a complete set of model forms corresponding to its normal scope of work, which can be readily adapted to the needs of each contract. This is the preferred arrangement as it enables the construction contractor's personnel to use documents with which they are familiar. A set of ITR forms covering the contract scope can be presented for approval with the ITP. Alternatively, the EPCM contractor may impose its own versions of the ITRs. Figure 5.3 presents a blank model ITR form upon which can be entered details of the project and materiel to be inspected, identifying the EPCM contractor, owner and construction contractor, citing the reference document(s), and giving space for the inspection and test points to be listed, allowing those which are hold points to be

EPCM contractor Name/Logo	**INSPECTION AND TEST RECORD** **(Indicate type of materiel)** Record no.	Owner: .... Project no. .... Project: ....

Construction contractor: .... Contract no. .... Contract:	Equipment: .... Tag no. .... Supplier: .... P.O. no. ....

Reference document(s): .... ITP no. ....

No.	Inspection/Test point	Constr. contr. inspected Initials/Date	EPCM contr.	
			Hold-point	Witnessed initials/Date

Remarks and conclusions

.............pages attached

For construction contractor	**For EPCM contractor**	**For ....**
Conforms to specified requirements	Reviewed/Accepted	....
Name:...	Name:..........................	Name:..........................
Signature:	Signature:	Signature: ....
Date:	Date:	Date: ....

Figure 5.3 Inspection and test record.

indicated. As the inspection and test points progress, they can be signed off. At the foot of the page there is space for comments and signatures to sign off and close out the form.

The time to start preparing the ITRs is as soon as possible after contract award. For example, if a construction contractor possesses a model form for packaged boilers of which there are, say, three to be installed on site, it shall review the form against contract requirements and adapt and modify the inspection and test points accordingly. A separate copy of the form shall be filled in with basic information including designation, tag number and reference documents for each of the three packaged boilers and inserted in the construction quality file (see Section 5.8) until required for recording inspection and test results. The same thing shall be done as far as possible for all materiel in all disciplines. At the same time those inspection and test points which are chosen as hold points shall be identified with the letter 'H' in column 4 of the ITR form. Thus, when the time comes for an inspection or test, preparation is done and the form is available for immediate use. This will save a lot of time later and thus contributes to the smooth running of the inspection and test programme.

Some ITRs will come from other parties, such as surveyors, laboratories, concrete batching plants and suppliers and will in each case have a format specific to the organization concerned.

5.8 Construction quality file

Accumulating quality records as the works advance is not enough. If they are to become complete and accessible, they have to be assembled physically in a logical coherent manner. We therefore require each construction contractor to establish and continuously update a construction quality file (CQF). The EPCM contractor will find it advantageous to impose on all construction contractors a standardized section title and numbering system. This can be taken a step further by the

EPCM contractor supplying as free issue to each construction contractor suitable folders and section separators. This canalizes the work of those responsible for the CQFs and facilitates the task of the EPCM contractor in monitoring the compilation of the documents and subsequently in reviewing and accepting them. It will also, during the future life of the plant, facilitate the task of anyone searching the CQF for a particular document. The following is a specimen outline list of contents to be adapted to the needs of a specific project.

A Site Quality Plan

B Materiel Procured by the Construction Contractor
For all significant materiel purchased, sufficient documentation for future operation, maintenance and repair, which includes:
- Unpriced copies of purchase orders
- Material certificates
- Suppliers' instruction manuals

C Specifications, Method Statements and Qualification Records
For welding, these include all applicable:
- Weld procedure specifications
- Procedure qualification records
- Heat treatment procedures
- Nondestructive examination (NDE) procedures

This section could also include other significant procedures, for operations such as:
- Heavy lifts
- Tank hydro tests and settlement measurements
- Application of internal coatings for tanks and vessels
- Application of refractory materials

D Personnel Qualification Records
Lists and copies of qualification certificates of:
- Welders
- NDE operators and inspectors
- Electricians

E Inspection and Test Plans, and Inspection and Test Records
A selection can be made from the list of documents cited hereunder corresponding to the discipline and scope of work of the contract, retaining the standard numbering for the sections chosen as appropriate for a given contract.

E1 General
- Inspection and test plans
- Inspection, test and measuring equipment calibration records

E2 Site Preparation and Earthworks
- Setting-out survey reports
- Soil bearing test reports
- Final survey reports

E3 Piling
- For bored piles, inspection and test records (ITRs)
- For driven piles, ITRs both for prefabrication and for site works
- Survey reports, out-of-position piles identified with corrective actions/concessions referenced back to nonconformance reports

E4 Concrete Batching Plant
- Initial survey and audit report
- Table of approved mix designs
- Certificates of approved sources of course and fine aggregate, cement, additives and water
- Test results of approved mixes

E5 Reinforced/Pre-stressed Concrete Works, *In situ* and Prefabricated
- Reinforcing bar material certificates
- Tensioning records
- Pre-pour inspection records
- Pour records
- Cube/cylinder crushing test records
- Curing and final inspection records
- Survey reports

E6 Building Works
- Foundation records
- Main structure records
- Roofing records
- Cladding and insulation records
- Building services records
- Finishing trade records

E7 Underground Piping
- Prefabrication records (see E15)
- Compaction test records for bedding and backfill
- Pipe site welding/coating records
- Pipe, fittings, manholes, etc., lay and joint records
- Hydrostatic test records

E8 Roads and Parking Areas
- Material certificates
- Asphalt batching plant records
- Bearing test records
- Finishing, curbing, drains and marking inspection records
- Survey reports

E9 Structural Steelwork Prefabrication
- Material certificates
- Weld inspection records
- NDE records
- Destructive test records
- Surface preparation and paint records
- Dimensional inspection records
- Weld maps for traceability
- Final inspection records prior to shipment

E10 Structural Steel Erection
- Grouting and bolting down records
- Bolt torque records
- Structure level/plumb survey records
- Site welding records

E11 Site Erected Storage Tanks
- Foundations records (see E2 and E8)
- Prefabrication records (see E9)
- Site welding records
- Site NDE records
- Site destructive testing records
- Site dimensional check records
- Site vacuum box test records
- Hydrostatic test/settlement measurement records
- Site surface preparation coating/painting records

E12 Static and Rotating Equipment Erection
- Level, verticality, positioning and orientation inspection records
- Grouting-in and bolting-down records
- Internals and accessories installation records
- Pressure test records
- Shaft alignment records
- Auxiliary systems check records
- Preservation and maintenance records

E13 Lifting Devices
- Assembly inspection records
- Electrical test records
- Safety devices and limit switch test records.
- No-load-test records
- Safe working load and overload test records.

E14 Refractories
- Qualification test records
- Periodic sampling test records
- Repair records
- Stage and final inspection records

E15 Piping Prefabrication
- Material certificates
- Welding inspection records
- NDE records
- Destructive test records

- Heat treatment records
- Dimensional inspection records
- Surface preparation and painting records
- Weld maps for purposes of traceability
- Final inspection records prior to shipment

E16 Piping Erection (a seamless continuation of records established during prefabrication)
- Welding inspection records
- NDE records
- Destructive test records
- Heat treatment records
- Pipe support records
- Flange bolt torque records
- Overall conformity to piping and instrumentation diagrams (P&ID) records
- Flushing/air-blowing/reinstatement records
- Hydrostatic test records
- Surface preparation and painting records
- Weld maps for purposes of traceability

E17 Electrical Installation
- Equipment installation inspection records
- Cable support/protection inspection records
- Cabling ITRs
- Earthing ITRs
- Equipment test/relay setting records
- Lighting and low power circuits ITRs

E18 Instrumentation and Control
- Field instrument, control valves and relief valves pre-installation inspection and test records and calibration records
- Field instrument, control valves and relief valves *in situ* installation/connection inspection and test records
- Cabinet and console installation inspection records
- Cabling ITRs
- Earthing ITRs
- Piping systems ITRs
- Loop test records

E20 Painting
- Material certificates
- ITRs

E21 Insulation
- Material certificates
- Inspection records

F Nonconformity Reports (NCRs)
Lists and copies of all NCRs relevant to the contract each complete with supporting documents

G 'As-built' Information.
A complete list of EPCM contractor's and construction contractor's drawings distinguishing between:
- Those drawings marked up in red to show site modified 'as-built'
- Those drawings already in effect 'as-built' without modification

It may be more practical to deliver the 'marked-up' drawings separately rather than physically including them in the CQF (see Section 6.9).

H Incomplete Works
A list of works included in the contract but which have not been for valid reasons completed (see Sections 11.4–11.6 and Figure 11.2).

How many copies of the CQF are needed? The answer is as few as possible consistent with project needs. An excessive number of copies turns the site office into a printing shop and more importantly brings the system into disrepute. The absolute minimum is a single CQF made up of original documents, but it is safer to establish a second CQF copy in another location as a back-up. Very special attention must be focussed on documents required as part of the authorities approval process. The best way to ensure that a construction contractor is motivated to produce quality records and to continue to build up the CQF is to subject payment for works completed to the presentation of the corresponding quality records.

Documents in hard copy are needed for the day-to-day quality management of the site and also for the question of authentic signatures. If documents are to be scanned to present the CQF in electronic format, the necessary software/ database must be provided in good time.

5.9 Inspection, measuring and test equipment

For results to be valid and be perceived as such, it is essential that all parties have confidence in the accuracy of the inspection, measuring and test equipment used. Those directly concerned are in the first place the construction contractors and special-service providers, but there may well be others. This implies that their quality plans include provision for identifying and tagging all such equipment, defining the types, methods and frequencies of calibration required, either at timed intervals or immediately before use, and setting realistic limits of acceptable tolerances. The whole programme must be duly recorded on suitable documents, with responsibilities clearly defined for setting up and for ongoing implementation, to be audited at regular intervals. Items to be included for regular checking and calibration can include:

- Levels and theodolites
- Weighing equipment
- Pressure gauges
- Laboratory and test bay equipment
- Electrical test apparatus

CASE STUDY

On a process plant construction site there were 19 principal construction contractors. Each contract had been awarded on a competitive tendering basis from a short pre-selected list of bidders. On the whole, the contracts ran smoothly, any claims for extra payments being resolved on an equitable basis. As part of the end of project report, the quality of the works completed by each contractor was assessed. The overall ratings for quality of the works were as follows:

- Excellent 4 firms
- Good 7 firms
- Average 5 firms
- Poor 2 firms
- Unacceptable 1 firm

Concerning the 'Unacceptable' works, which were performed early in the construction phase, this was due in large part to poor performance on the part of the management and supervisory staff the construction contractor had sent to site. Not surprisingly, the contractor performed badly in other areas, but some of the blame must go to the EPCM constructor whose own supervisory staff were at the time under strength. The contract was terminated, remedial works being undertaken by others. Both companies whose works were rated as 'Poor' had inadequate support from head office, mediocre management on site and not enough of their own skilled personnel to stiffen up their teams made up of a very large majority of temporary workers. Firms with an 'Excellent' rating gave good support to their site, assigned high-quality managers and supervisors and drafted a solid core of their own long serving skilled workers. In general, performance in other domains, such as safety and respect of the schedule, was also excellent.

Lessons learnt

- The main factor affecting quality of the works which emerged was the human one, that is, head office support, site management and supervision. A too large proportion of temporary staff could have a detrimental effect.
- This needs to be taken into account during bid clarifications and contractor selection.

6

Construction–Engineering Interface

6.1 Introduction

The EPCM contractor's engineering department (ED) will have been responsible for designing the plant and for the technical input to purchase orders and construction contracts. These tasks will not have been fully completed at the time the site is opened. The ED will now assume the additional responsibility of providing the EPCM contractor's site team with the following inputs:

- Furnishing drawings and other documents 'approved for construction'
- Answering site-generated engineering queries
- When requested, defining the resolution required for site nonconformities
- Providing technical liaison with utility companies, the insurer and authorities
- Generally providing engineering support to the construction effort
- Overseeing site modifications and their recording on 'as-built' drawings
- Giving support for and participating in final inspection and testing leading to 'construction completion' and 'turnover' to the owner

So where should the interface be between construction and engineering? The availability of electronic means of communication and transmission of documents certainly facilitates exchanges between construction personnel on site and the head office ED. However, it is preferable, even essential, for the ED to be represented on site, by at least one person. Let us call him or her the **site engineer**, to be the focus and the communication channel for all engineering matters. Depending on the size of the project there could be, not just one individual, but a team as suggested in the typical site organization chart presented in Figure 4.1.

6.2 Types and origins of engineering documents

The following classes of engineering documents will be required on site during the construction phase:

- EPCM contractor's specifications and drawings
- Suppliers' documentation corresponding to materiel furnished
- Construction contractors' specifications and drawings (When an element of design is included in the contract scope)
- Standards and codes of practice

An order of precedence between the different types of document should have been defined early in the project to be applied to any inconsistency which may come to light. Each of these categories of documents is discussed in more detail hereunder. These documents may be in hardcopy and/or electronic format to meet project requirements, but hardcopy documents are required at the work face by those directly involved in construction, erection, inspection and testing.

6.3 EPCM contractor's specifications and drawings

These documents will have been created in the EPCM contractor's engineering department (ED) from the starting point of the basic design provided by the owner. They will have gone through the iterative process of review and comment within the ED and with other parties involved, in particular with the owner and where applicable with the authorities concerned. Some of them will have been included, at least in preliminary form, in the bid packages and then as part of the construction contracts. They will be issued to the EPCM constructor's site team at 'approved for construction' (AFC) status, for forward issuing to the construction contractors for execution. There may be cases where preliminary versions of drawings may be sent to site for information, but this should be the exception and their non-AFC status should be boldly indicated.

There will be a number of basic specifications for each discipline, defining the ground rules, and then specifications applicable to specific works. It is important that the extent of applicability of the basic specifications to each construction contract be defined as soon as possible. If only a very small part of a given basic specification is applicable to a given contract, it should be clearly stated.

6.4 Suppliers' documentation

A purchase order (PO) from the EPCM contractor is the basis on which a supplier will provide materiel. The PO will require the supplier to generate a certain number of documents. These can include:

- General arrangement drawings including weights, and lifting details
- Drawings of anchor bolt arrangements
- Drawings of electrical and piping connections, that is, flange and terminal box details
- Electrical, hydraulic and pneumatic schematics and connection diagrams
- Data sheets for components
- Unpriced copies of suborders
- Material and test certificates for components
- Spare-parts lists

- Records of inspection and tests carried out before shipment
- Certificates issued by an authority in the country of manufacture
- Packing and shipping instructions
- Erection instructions
- Lubrication charts
- Site inspection, test, start-up and commissioning instructions
- Operating and maintenance instructions

Certain of these documents will be transmitted to the EPCM contractor for review and eventual approval before manufacture begins. When a supplier document once approved replaces an EPCM contractor's document, there must be a mechanism to formalize this change. For example, should the EPCM contractor create a preliminary schematic diagram for an electrical switchboard in order to define requirements, and the supplier from this starting point produces and manufactures according to its own definitive diagram duly approved by the EPCM contractor, it is the latter diagram which must be used on site during erection and testing. It is important that any modifications carried out on materiel to correct anomalies detected during shop inspection and testing are incorporated into the corresponding supplier documents to be raised to 'as manufactured' status.

Other supplier documents will be verified at source by the EPCM contractor's inspectors. The PO will also define the number of copies and the presentation, that is, hardcopy and/or electronic format. The documents from each PO will be assembled into a single volume or a set of volumes called 'manufacturing data book' (MDB) or something similar. It is advantageous that the format and presentation of these documents be standardized across all of the POs in order to facilitate their use. The MDB should be reviewed by the inspector before dispatch.

In an ideal world the MDB would be available at the time that materiel is shipped, but in practice too often its assembly drags on, with the various documents scattered throughout the supplier's premises. Even when it is finished, it may not be where it is wanted on site. Linking a substantial payment to the delivery of the MDB should speed up delivery. Another approach would be to make special arrangements to ensure delivery of copies of those documents required early on-site for erection, including, in particular, general arrangement drawings and anchor bolt drawings necessary for the associated civil works.

6.5 Construction contractors' specifications and drawings

For some contracts of a specialized nature, the scope of work of the construction contractor may include a design element giving rise to specifications and drawings. This could be the case for a cooling tower or a water treatment plant for example. These will go through a review and approval sequence with the EPCM contractor. They will be issued to site at 'approved for construction' status. When a construction contractor's document supersedes an initial EPCM contractor document, the change needs to be formalized.

6.6 Engineering standards and codes of practice

Engineering standards, including codes of practice, are the basis on which a process plant is designed, specified and built. They act as a common currency between the involved parties and enable full use to be made of accumulated knowledge and wisdom.

They can originate from international bodies such as:

- International Organization for Standardization (ISO)
- International Electrotechnical Commission (IEC)

There is a large range of standards published by or under the auspices of national bodies such as:

- British Standards Institution (BSI)
- Association Française de Normalisation (AFNOR)

- Deutsches Institut für Normung (DIN)
- American National Standards Institute (ANSI)

The publications of several American private organizations are often specified on process plant projects worldwide, the main ones including:

- American Society of Mechanical Engineers (ASME)
- American Society for Testing and Materials (ASTM)
- American Petroleum Institute (API)
- American Welding Society (AWS).

On some process plant projects owners will be applying their own standards and codes of practice. These can be significant and their implications must be understood and taken into account.

Then there are also various standards and guides published by bodies of manufacturers, for example:

- Tubular Exchanger Manufacturers Association (TEMA)
- Conveyor Equipment Manufacturers Association (CEMA)

The applicable standards are selected at the basic design stage so as to permeate detailed design and procurement. By the time construction is at hand, the requirements of the selected standards are already embedded in the documents received on-site 'approved for construction'. However, there are some areas where the standards affect directly the site construction, erection, inspection and testing processes. These must be identified as soon as possible and access to them made available to those on site.

Unfortunately, on some projects, there is a tendency in specifications to state that the materiel or works must comply with a multitude of standards, the objective being to cover all foreseeable possibilities, putting the onus on the supplier or construction contractor to ensure compliance. At the bid stage, the potential supplier or construction contractor may well simply ignore these requirements, or might not have the time to take them into account, and furthermore the exact application of the standard(s) may not be resolved during bid clarification. As a result an ambiguous situation can develop on site. The only solution is to grasp the nettle as soon as possible to clarify the situation, defining exactly not only which standards apply, but which specific sections of the documents are concerned. If a standard has, say, 200 pages, but only the content of a few of the pages applies, let us identify them as soon as possible.

6.7 Document control

The EPCM contractor's document control system will be operative from start to finish embracing all phases of the project. The site document control system will be as it were an extension of the over all project system. The latter normally includes the following features:

- Alphanumeric document identification encoded with such information as discipline/system/process unit/off-sites/infrastructure, and so on
- Master lists of EPCM contractor's documents including drawings, specifications and requisitions, initially as lists of documents to be produced, subsequently recording generation of documents and successive revisions status
- Similar lists for suppliers' and construction contractors' drawings and specifications provided contractually and integrated into the project alphanumeric identification system
- Registers listing equipment, pipe lines, cables, valves, instruments, ans so forth, with salient information indicated for each item
- A dispatch facility for issuing documents to the various interested parties, recording all transactions made
- An archive covering all documents past and present

The site document control function is sometimes treated as an administrative matter, but it is much better if it is under the authority of the site engineer. This function needs the necessary procedures, staff, computing facilities and suitably equipped and furnished offices. Responsibilities are essentially:

- The receipt and registering into the site system of documents coming from all sources
- Issuing documents both within the EPCM contractor's site organization and to the construction contractors
- Ensuring that out-of-date revisions of documents are withdrawn
- Establishing and updating records of all the above transactions
- Archiving for reference all revisions of all documents received

Having engineering documents on site is of little use if they are not readily accessible to authorized persons. There is thus considerable advantage to be gained in establishing and maintaining an open library space with adequate viewing facilities, both in terms of computing terminals and large sloping desks where drawings can be spread out. Regarding electronic transmission of documents between head office and site, see Section 6.10.

6.8 Engineering site queries

A member of the EPCM contractor's site team, the owner's representative or a construction contractor, in fact any concerned party, may initiate a query concerning the on-site interpretation or application of an engineering document. The query could typically concern:

- A proposed substitution of material
- The resolution of a clash between two disciplines
- The difficulty of applying specified requirements to site conditions
- A mistake, real or suspected, in an engineering drawing already at 'approved for construction' status

It is important that each query is documented and appropriately processed. In Figures 6.1 and 6.2 there are

<table>
<tr><td>EPCM contractor

Name/Logo</td><td colspan="2"><u>SITE ENGINEERING QUERY RECORD</u></td><td>Report
No. ....</td></tr>
<tr><td colspan="2">Owner: ....
Project no. ....
Project: ....</td><td colspan="2">Contractor/Supplier: ....
Contract/P.O. no. ....
Contract/P.O. : ....</td></tr>
<tr><td colspan="4"><u>A. QUERY</u> (Eventually with proposed solution)
Reference document:

<u>Originator</u>
Representing: ....
Name: ....
Signature: ....
Appendices A1 to A .... attached Date: ....</td></tr>
<tr><td colspan="4"><u>B. REPLY</u> (To be decided by site engineer, alone or after consultation with engineering department or others. Copies of relevant correspondence to be attached)

<u>Site engineer</u>
Name: ....
Signature: ....
Date: ....
Appendices B1 to B .... attached</td></tr>
<tr><td colspan="4"><u>C. CLOSE-OUT</u>
The solution presented in Section B has been actioned, so this site engineering query is now closed out.

<u>Representing: ....</u> <u>Site engineer</u>
Name: .... Name: ....
Signature: .... Signature: ....
Date: .... Date: ....

Appendices C1 to C............. attached</td></tr>
<tr><td colspan="4"><u>Distribution:</u></td></tr>
</table>

Figure 6.1 Site engineering query record.

EPCM contractor Name/Logo	SITE ENGINEERING QUERY LOG AND STATUS REPORT Updated: ...../...../..... Page: ....../......					Owner: .... Project no. .... Project: ....
SEQ. no.	Originator	Discipline/ location	Definition of query	Resolution retained	Date established	Date closed out
Distribution:						

Figure 6.2 Site engineering query log and status report.

two forms which together are the basis of a procedure to manage site-generated engineering queries. It works as follows:

- The originator uses Part A of the form proposed in Figure 6.1, site engineering query record (SEQR), to define the query and where applicable propose a solution, adding to the form any attachments he or she may consider necessary to illustrate the query; he or she presents the form and attachments to the EPCM contractor site engineer.
- The latter shall number the form and register the query into the site engineering query log and status report (SEQLSR), a model of which is shown in Figure 6.2.
- The site engineer is either able to answer the query himself or herself or has to refer it to the EPCM contractor's head office engineering department.
- Once it is defined, the site engineer writes the answer/solution on Part B, with attachments if necessary, and re-issues the form to the originator for information and where applicable as an instruction.
- The originator receiving the instruction executes it and confirms in Part C that the instruction has been followed.
- The query is closed out by the counter-signature of the site engineer in Part C.
- The site engineer tracks the status of all engineering site queries through to close-out by means of the SEQLSR.

Some overlap may be perceived between, on the one hand, an 'engineering query' described above and a 'nonconformity' discussed in Chapter 8. The dividing line between the two concepts is that should an anomaly be identified before the works have started, it can be processed as an engineering query, and if detected once the works in question have been started or even completed, then it can be treated as a nonconformity. Borderline cases must be determined on a common sense basis.

6.9 'As-built' drawings

Right at the beginning of a project a policy decision must be taken to decide:

- What type of 'as-built' information must be recorded
- By whom, how and when
- At what point and in which manner is the information to be consolidated and eventually handed over to the owner

As a minimum, site-generated changes to underground works such as piling, drains and buried cables as well as to electrical and instrument connections and other items affecting safety or operations must be recorded. Primary responsibility for this can best be included in the contracts awarded to construction contractors, but this needs very close on-going proactive supervision on the part of the EPCM contractor, in the person of its site engineer or one of his or her assistants. It is important that a specific list be established of drawings that are eventually to be raised to 'as-built' status. Input information can be derived from engineering site query records and nonconformity reports, but most of all as the works progress by going out on site to check what has been done against the drawings and marking-up the latter in red to record any divergences.

When does the operation stop? Whilst there will be 'as-built' information generated during construction, it is, above all, during the commissioning and start-up phases that anomalies become apparent and modifications are made. At some predetermined point the marked-up information must be inputted into the drawing data base to raise the drawings therein to 'as-built' status. This should be done by the staff of the site engineer rather than in the head office, as it permits any queries to be clarified by a visit *in situ*. The same principle applies to construction contractors' drawings. It is important to establish as soon as possible when and how 'as-built' information is transferred to the owner, who will from then on be responsible for any further updating (see Section 11.3).

6.10 Information technology infrastructure

In recent years there has been a rapid development of progressively more and more powerful information technology (IT) combined with increased capacity to transmit information over telecommunication channels. This has transformed the manner in which information is created, stored, accessed, processed and transmitted. As a result process plant projects today can be organized around an integrated, interactive information management system, where operations such as approvals, revisions, issues and transmission can take place without handling hard copy, that is to say, a paper-free system. However, there is a gap, slowly narrowing, it is true, but nevertheless present, between what the IT industry can offer and what can be made to work simply, effectively and reliably on the average process plant project, in particular on the construction site.

At head office level, the information system could span the EPCM contractor, the owner and possibly some of the major suppliers and construction contractors depending on the degree to which management of the project is integrated. Typically the EPCM could create and manage a server, which would host the documents prepared by all of the participants, each of whom could update its own data base from that of the server. Alternatively, for a smaller project or where none of the partners possessed the necessary level of IT capability, the task of running the server could be undertaken by a special service provider. Either way there would be IT professionals monitoring the system, who would be able to intervene in case a problem occurred.

However, with the opening of the construction phase, the successful extension of the project information management system to the EPCM contractor's office on the site where the plant is to be built will depend on the following requirements being met:

- There has to be a secure, dependable telecommunication link of sufficient capacity between head office and site (depending on the geographical location of the site, this could be difficult to obtain).

- The site director must be IT competent, enthusiastic and must want to make the system work.
- There has to be a resourceful IT technician on site, who should be able to solve rapidly on his own any problems which may arise, this being especially vital for a site in a remote location.
- All future users based on site need to receive appropriate and sufficient training.
- Time is of the essence; so the system on site must be ready to be set up and rendered operational within an acceptably short time.

Most important of all is that the system hardware and software be simple, robust and proven, with all the anomalies already expunged. A critical construction site with a very tight time scale is no place to resolve the teething troubles of a new untried system. It is better by far to have a basic system that works rather than one with the potential to provide a more extensive range of functions, but which is unreliable. Depending on circumstances, if the telecommunications link were unsuitable, there could be occasions when it would be advantageous to send by courier CDs on which are registered large numbers of documents.

In defining the requirements for the extension of the project information system to the site, the following questions need to be addressed:

- What are the documents present in the main project database that can be accessed, viewed, downloaded and printed by EPCM personnel on site (e.g. specifications, drawings, equipment lists, data sheets, etc.)?
- Do these documents only become available when they have reached a certain status (e.g. 'approved for construction')?
- What is the mechanism for alerting site as new documents become available, or existing ones are modified?
- Which documents present in the project database, if any, can be modified from site (Perhaps only to raise them to 'as-built' status)?

- What are the restrictions to be placed on individuals for access to documents on a 'need to know' basis and what is the password system to enforce them?
- What are the templates to be used for exchanges between head office and site as part of formal procedures (e.g. for processing engineering queries or nonconformities)?

The latter point is essential as unrestricted email exchanges can lead to formal procedures being bypassed and devalued, with all the potential dangers that that can represent.

The persons carrying out the physical work of building the plant will require hard copies of specifications and drawings. The question then arises as to when and where the transition from electronic format to hard copy is made. Whilst it can be desirable for construction contractors on site to be able to screen view documents relevant to their respective contracts, do we expect each of them, even the smaller ones, to invest in the equipment to print out the larger-size drawings? Do they have on site sufficient IT capability to participate fully in the EPCM contractor's IT structure and resolve any problems which might arise? Might it not be too complicated? There is a strong case for each new batch of 'approved for construction' drawings and other documents to be printed out by the EPCM contractor on site and to be issued as hard copy in a traditional manner under cover of a transmittal note. This could give better control, both technically and contractually, especially as at the same time superseded revisions can be removed.

At the limit there could be circumstances, for example, on a small site situated close to head office, where it might be cheaper and simpler to send packets of documents to site by post or by courier rather than getting bogged down in the provision of an IT system disproportionately complicated for the size of the job.

It is evident that the IT plays a vital role in the over all site quality management system, so it needs to be continuously monitored and subjected to periodic audit.

CASE STUDY

On a process plant construction site, numerous items of equipment were to be positioned on several floors of a steel-framed building. Piping and electric and instrumentation and control cabling were close-packed with numerous wall and floor penetrations. The schedule was such that prefabrication of structural steel had to be launched before all equipment details were fixed. Inevitably, piping and cabling were affected. This gave rise to a large number of clashes leading to modifications, which were taking a long time to resolve, as there was not a strong engineering presence on site and the EPCM contractor's head office engineering department was becoming involved in the next project. Although the construction contractors' managements obviously perceived the modifications as an occasion to boost their profits on the project, their execution and supervisory staff were to say the least demoralized.

The situation was eventually improved by sending design engineers and draughtsmen to site to decide changes on the spot with a streamlined procedure for engineering review and updating of drawings, all tied in with a rapid change order system.

Lessons learnt

- This project may have been reaching its limits of 'schedule compressibility'.
- When prefabrication and construction have to proceed with many design details not yet fixed, extra space must be allowed, even at the cost of a slightly bigger building. Bigger wall and floor openings, and more of them, can be incorporated into the original design at little or no extra cost, to give flexibility in case piping and cable runs have to be rerouted. Excessive openings can always be filled in afterwards.
- Given such a situation, the persons who design the plant can deal with engineering problems better and more rapidly if they are based on site.

7

Materiel Management

7.1 Sources of materiel

Depending on the contractual strategy decided for the project (see Section 1.4), materiel received by the construction contractors for incorporation into the plant may come from several different sources as listed hereunder:

- Procured by the EPCM contractor head office, then issued free to the construction contractors
- Procured by the EPCM contractor site office, then issued free to the construction contractors
- Free issued by the owner to the EPCM contractor, then in turn issued free to the construction contractors.
- Procured directly by the construction contractors

Major items of equipment are normally procured by the EPCM contractor's head office. The EPCM contractor's procurement from the site office is generally limited to minor items purchased locally for speed and convenience. Items issued free by the owner may be special process equipment. For civil works and buildings, the construction contractors normally procure items such as pre-mixed concrete, reinforcing bars, doors, windows, and so forth. Concerning piping contracts, pipes, valves and fittings may be procured either by the

EPCM contractor or by the construction contractor according to the strategy retained. The situation is similar for electric cables, fittings, cable racks and trays. In the specific case of a construction contractor designing and building a complete self-contained facility, it will probably procure the complete range of materiel necessary.

Effective management of materiel on site is an essential element of mastering quality. This chapter is focussed on materiel management exercised by the EPCM contractor, but the principles evoked apply equally to the construction contractors. Figure 7.1 is a materiel flowchart outlining the passage of materiel from the different sources to its incorporation in the works.

7.2 Purchase orders

Purchase orders (POs) can be broadly classified into the following categories for the supply of:

- Items of equipment such as pressure vessels, transformers or pump sets
- A range of material for a given discipline such as piping, electricity or instrumentation
- All materiel for a multidiscipline ensemble such as a major turbo-compressor unit, including control panel and auxiliaries, virtually a self-contained project within the project

In an ideal world, POs are placed with competent suppliers, each operating an effective quality management system to ensure that the materiel conforms to the specifications, is inspected and tested, and is properly packed and treated with care during shipment. On arrival on site, materiel can then be readily incorporated into the works with no problems. Unfortunately, reality sometimes falls short of this ideal, and so a procurement problem becomes a site problem. Engineering follow-up, inspection at source, expediting and shipping are outside the scope of this handbook, but it is important that all parties recognize that attention paid to procurement pays

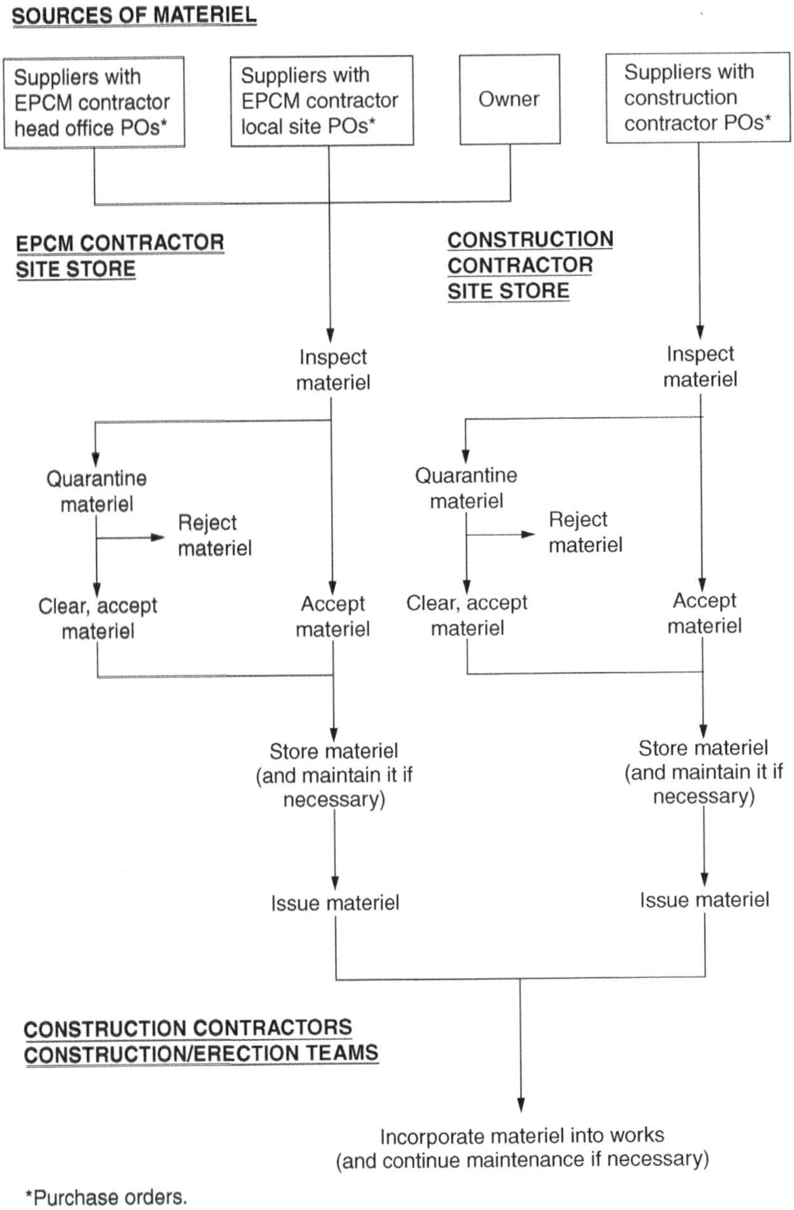

Figure 7.1 Materiel flowchart.

dividends in problems avoided on site. In particular, close care and attention must be directed on the following areas before materiel is shipped:

- *Interfaces* – anchor bolt arrangements, pipe flange connections, cable glands and cable terminals
- *Trial erections and dimensional checks* – making sure items will fit together
- *Functional and performance tests* – making sure everything works
- *System tests by hooking up the constituent items* – checking compatibility and functionality
- *Labels and tags* – making sure that all items are correctly identified
- *Authorities* – meeting the requirements in the country of manufacture and the requirements of authorities of the country in which the plant is being built as well as those of the utility companies and the insurer
- *Surface preparation and painting* – properly carried out and not a rush job just before shipment
- *Packing* – taking into account the form of transport, the climatic conditions and possible long delays waiting for customs clearance
- *Packing lists* – complete, accurate and tied back to the items of the purchase order to facilitate reception and identification on site
- *'As manufactured' drawings, lubrication schedules, erection, commissioning, start up, operating and maintenance instructions and spare parts lists* – delivered to the site on time when they are needed
- *Spare parts for commissioning and start-up* – ordered in time so that they can be delivered with the equipment
- *Sub-contractors of a supplier* – need to be closely monitored too

Over all, problems are much more easily dealt with at the supplier's before shipment than after arrival on site. This is especially true when the site is in a remote area with little in the way of industrial infrastructure. However, this does not preclude the shipment to site of a major item of equipment, vital to maintain the construction schedule, but with a minor component missing, which can easily be fitted later, such as a missing pressure gauge on a pump. In these circumstances the decision to ship must be taken by the engineering department and not left to the discretion of an expeditor.

7.3 Materiel storage facilities

The EPCM contractor needs to build a warehouse and enclose, drain and surface an outdoor storage yard of adequate dimensions to accommodate the types, volume and rate of throughput of the expected materiel. The EPCM contractor will possess a detailed procedure defining the type of storage for each category of materiel, either in the open, under cover, within a building under ambient conditions or in an air-conditioned space. The facility shall include an area for receiving and inspecting incoming goods and a quarantine zone. A grid reference system shall be established throughout the warehouse and yard to identify the geographical position of each item stored. For reasons of size and/or weight and to avoid double handling, some items of equipment may be delivered directly to their place of installation. Another case where equipment is delivered directly concerns delicate items, in particular computing equipment, to be installed in a finished building, with its air-conditioning system already in operation. In these cases the equipment should still be submitted to the normal administrative procedure for receiving and processing incoming materiel, followed by its formal issue to the construction contractor concerned, even though the equipment has not physically passed through the warehouse.

7.4 Materiel control procedure

The EPCM contractor normally possesses a corporate materiel control procedure, which can be adapted to each specific project. Items are listed at the requisition stage with an alphanumeric identification, and then followed through procurement, release for shipment, shipment, arrival on site, then issued for construction. Materiel control on site is thus a facet of a larger system covering the whole project. Electronic exchange of information between head office and site is discussed in Section 6.10. Ideally the related software should be linked to that used for material take off, particularly in the piping discipline, so that picking lists derived from piping isometric drawings can be compared with lists of piping material available on site. Thus, priorities and shortfalls can be identified. As far as the site is concerned, the EPCM contractor's materiel control section's activities embrace the following:

- Incoming inspection of materiel for conformity to purchase order requirements in terms of quantities, marking, traceability to accompanying documents and absence of damage.
- Storage of materiel, with, where necessary, regular ongoing documented inspection and preventative maintenance.
- Issue of materiel to construction contractors correctly documented.
- Establishment and continual updating of records to enable details of materiel available on site to be known at any moment.

Once materiel has been issued to a construction contractor, the latter takes over responsibility for it up to the time of completion and turnover of the works. The construction contractor is equally responsible for materiel it has procured directly. This implies the creation and operation by the construction contractor of a materiel control system incorporating those features present in that of the EPCM contractor.

7.5 Incoming inspection, ongoing inspection and maintenance

Site materiel control procedures habitually incorporate incoming administrative inspection based on the well-known concept of an 'over, short and damaged (OS&D) report', which is exactly that. Any discrepancies are resolved by informing the EPCM contractor's head office procurement department, which will, if necessary, contact the suppliers and/or shippers as may be appropriate. It is highly desirable that this administrative inspection by the warehouse staff based on the establishment of OS&D reports be consolidated by an incoming technical inspection carried out by suitably qualified staff seconded on a part-time or full-time basis from the EPCM contractor's construction supervision team. Their objective will be to:

- Identify technical problems that may have slipped through the 'at source' inspection system run by the EPCM contractor's head office
- Identify those items requiring ongoing inspection and maintenance

The depth of incoming inspection can compensate for any shortcomings in the control exercised by the EPCM contractor's head office at the supplier's before shipment. In Section 7.2 we have already highlighted the importance of resolving materiel problems at the supplier's rather than on site. However, in view of the complexity and pace of a process plant project, in spite of all the care taken, some materiel may be delivered with physical or quality documentation problems, be they immediate or potential. So let us detect them as soon as the materiel arrives, so that corrective action can be launched without delay.

Another reason for incoming inspection is to identify where there is a need for ongoing inspection and maintenance. Items concerned included *inter alia*:

- Vessels under inert gas purge, whose pressure has to be regularly checked and if necessary topped up
- Electric motors whose shafts have to be regularly rotated by hand
- Equipment enclosures to be kept free of moisture by use of desiccant bags or anti-condensation heaters connected to a temporary electrical supply, with where appropriate periodic checks of insulation resistance

Once identified, the items concerned will be subjected to regular inspection and maintenance. A model procedure is proposed based on the use of three forms shown in Figures 7.2–7.4. This programme must take full account of the suppliers' instructions. Where appropriate, once equipment has been issued to the construction contractor, the latter becomes responsible for continuing the inspection and maintenance programme.

7.6 Traceability

Traceability is the mechanism by which it is possible to trace an element of the works from its final position on the site back through the various erection, prefabrication and manufacturing processes to its origins and to information concerning its characteristics. Its purpose is:

- to be able to prove that materials and processes used conform to specified requirements (to meet authorities' exigencies where applicable)
- in case a problem arises with work in progress or with the finished works, to enable other locations likely to experience the same difficulties to be identified, thus enabling the problem to be treated in its totality

The physical essence of traceability is the cross-referencing between documents and from documents to the corresponding materiel and works. This means a traceability trail has to be established that can be followed linking all elements concerned. The level of traceability required must be defined as early as possible in writing and must be a contractual requirement in the purchase orders and construction contracts concerned. Special care must be taken to ensure that authority traceability exigencies are incorporated right from the start. Specified requirements should not be exceeded. Excessive zeal in this direction is unnecessary, wastes money, can create frustration and diminishes the credibility of the mandatory requirements for traceability.

<table>
<tr><td>EPCM contractor

Name/Logo</td><td colspan="3"><u>SITE INCOMING INSPECTION RECORD
EQUIPMENT</u></td><td>Record
No. ....</td></tr>
<tr><td colspan="2">Owner: ....
Project no. ....
Project: ....</td><td colspan="3">Supplier: ....
P.O. no. ....
P.O. : ....</td></tr>
<tr><td colspan="5">P.O. item no. .... Shipping reference: ....
Equipment: .... Tag no. ....
Reference documents: ....</td></tr>
<tr><td colspan="2" rowspan="2"><u>Inspection point</u></td><td colspan="2">Acceptable</td><td rowspan="2"><u>Remarks</u>
(Action required)</td></tr>
<tr><td><u>Yes</u></td><td><u>No</u></td></tr>
<tr><td colspan="2">Transport damage

Name/data plate

Anchor bolt hole configuration, anchor bolts.

Centre lines marked

Pipe connections/nozzle orientations/ flanges/gaskets/screwed connections

Electrical/instrument junction boxes/ glands/earthing connections

Internals/externals/trial erection

Insulation, projections and supports

Painting state

Spare parts received

Erection, maintenance and operating instructions received</td><td></td><td></td><td rowspan="2"></td></tr>
<tr><td colspan="4">Type of storage required: outdoor/undercover/indoors/ air-conditioned*
Ongoing inspection and maintenance: inert gas purge/ shaft rotation/heating/dessicant bags/insulation resistance/other*
*Delete as necessary</td></tr>
<tr><td colspan="3"><u>Distribution</u></td><td colspan="2"><u>For EPCM contractor</u>
Name: ....
Signature: ....
Date: ....</td></tr>
</table>

Figure 7.2 Site incoming inspection record – equipment.

<table>
<tr><td colspan="2">EPCM contractor

Name/Logo</td><td colspan="3"><u>SITE INCOMING INSPECTION RECORD
MATERIAL</u></td><td>Record
No. ....</td></tr>
<tr><td colspan="3">Owner: ....
Project no. ....
Project: ....</td><td colspan="3">Supplier: ....
P.O. no. ....
P.O. : ....</td></tr>
<tr><td colspan="6">P.O. item no. .... Shipping reference: ....
Material: ....
Reference documents: ....</td></tr>
<tr><td colspan="2" rowspan="2"><u>Inspection point</u></td><td colspan="2">Acceptable</td><td colspan="2" rowspan="2"><u>Remarks</u>
(Action required)</td></tr>
<tr><td><u>Yes</u></td><td><u>No</u></td></tr>
<tr><td colspan="2">Transport damage</td><td></td><td></td><td colspan="2"></td></tr>
<tr><td colspan="2">Marking/tagging</td><td></td><td></td><td colspan="2"></td></tr>
<tr><td colspan="2">Mill/test certificates received</td><td></td><td></td><td colspan="2"></td></tr>
<tr><td colspan="2">Traceability material/documents</td><td></td><td></td><td colspan="2"></td></tr>
<tr><td colspan="2">Protection arrangements</td><td></td><td></td><td colspan="2"></td></tr>
<tr><td colspan="4">Type of storage required:
Outdoor/undercover/indoors*

Maintenance required, if any*

*Delete as necessary</td><td colspan="2"></td></tr>
<tr><td colspan="3"><u>Distribution</u></td><td colspan="3"><u>For EPCM contractor</u>
Name: ....
Signature: ....
Date: ....</td></tr>
</table>

Figure 7.3 Site incoming inspection record – material.

EPCM contractor Name/Logo	**SITE STORAGE INSPECTION AND MAINTENANCE RECORD**	Record No. ....

Owner: .... Project no. .... Project:	Supplier: .... P.O. no. .... P.O. : ....

Equipment: ....Tag no ....
Type of inspection/maintenance: Inert gas purge/shaft rotation/heaters/dessicant bags/insulation resistance check/other* .... Frequency: daily/weekly/monthly other* ....

*Delete as necessary

Planned date	Actual date	Acceptable		Remarks (Action required)
		Yes	No	

Distribution	**For EPCM contractor** Name: .... Signature: .... Date: ....

Figure 7.4 Site storage inspection and maintenance record.

Traceability requirements on site can vary widely from one project to another and can be required in any discipline. Appendices A and D discuss traceability respectively in the context of concrete and of welding.

7.7 Spare parts

At the beginning of a new project the owner may decide that for some items of materiel the types, models and manufacturers specified shall be similar to those present in its existing plants, as this will simplify subsequent maintenance, in particular concerning the management of spare parts, with considerable savings in cost. This is particularly true when the project is an extension on the same site of an existing plant, and even without such a precedent, there is a good case for standardizing as much as possible across the different contracts and purchase orders. Certain major long-lead items, such as say a spare rotor for a turbine, must be ordered at the same time as the rotor of the machine concerned, so they can be manufactured together. For other items a scale of provision must be decided, generally in terms of quantities required, on the one hand for commissioning and start-up and later for say the first 2 years of commercial operation of the plant. There is an obvious advantage in including spares in a supplier's scope from the beginning, to be delivered at the same time as the main materiel. On site they must be carefully stored and identified. At turnover, transfer of spare parts will be made from the EPCM contractor's site store to that of the owner.

CASE STUDY

On the construction site of a process plant, pipework prefabrication and site erection were already well advanced when a number of significant surface defects on piping material came to light, potentially putting into doubt the integrity and safety of the

Continued

CASE STUDY (Continued)

finished installation. A third party competent in metallurgy and nondestructive examination was engaged to inspect all known defective items, identifying the manufacturers and heat numbers concerned. Fortunately, from the very inception of the two piping contracts, an appropriate traceability system had been instigated and diligently applied. This enabled other elements of pipe, from the same heats as the defective ones, to be located and examined in the warehouse, in the prefabrication shops, in the blast and paint shop, in various buffer stocks and already erected on site. Defects were ground out and measurements of residual wall thickness were made on the ground-out areas to be compared with code requirements leading to rejection or acceptance. The whole operation was controlled and documented through the site nonconformity management system, which, in addition to its quality function, also enabled a strong case to be built up, leading to a financial settlement out of court. The problem was in fact limited to piping material from a small number of heats from two little-known manufacturers situated in a developing country, probably selected for their low prices. Before shipment to site only a rudimentary inspection had been made at the stockist through which the material transited.

Lessons learnt

- Procurement shortcomings, in this case poor selection of suppliers and inadequate inspection prior to shipment, can lead to a site problem.
- Maintaining traceability is a chore requiring constant application, but if you run into trouble you will be very glad that you have it.
- The formality and structure of a nonconformity management system enable a mass of detail to be marshalled to form a coherent record, which apart from its technical merit, can if required provide valuable evidence for resolving the question of contractual responsibility and damages.

8

Nonconformities

8.1 What are nonconformities and how do we manage them?

A **nonconformity** is deemed to exist when a significant deficiency in materiel, works, procedure, documentation or organization results in a condition not in conformity with specified requirements, and, moreover, which cannot be corrected using an existing procedure within the scope of normal supervision. The notion of nonconformity is not applied to works temporarily incomplete.

The dividing line between what is and what is not a nonconformity can be woolly, but the examples given in Table 8.1 illustrate the principles to be followed when deciding how to react to a given case. Borderline cases have to be resolved on a common sense basis.

Whether there is a formal system or not nonconformities manifest themselves, so one way or another they have to be processed. On a construction site without a formal system, information concerning nonconformities may be scattered about in various files or may be simply unrecorded or even covered up, and so later it is difficult, if not impossible, to identify the problems and to learn how and

Table 8.1 Examples of what is and what is not a nonconformity.

Example	This is a non-conformity because	This is not a non-conformity because
Concrete works		
Step change in concrete crushing strengths to below specified minimum value	Investigation needed to find root cause. Defective concrete zones to be identified, corrective actions/concessions to be defined.	
After removal of formwork, some cosmetic repair is necessary		Can be corrected with existing procedure.
Structural steelwork		
Major misalignment between prefabricated element and anchor bolts/foundation.	Engineering review needed. Do we change steelwork, concrete or both?	
Burrs not removed from drilled holes.		Can be removed following usual trade practice.
Mechanical equipment		
Overhead travelling crane fouls building structure.	Engineering review needed to determine necessary modification	
Shafts of motor/pump set misaligned.		Can be corrected following usual trade practice. May just be incomplete work.
Welding		
Repeated occurrence of heavy weld defects.	Investigation needed to find root cause. Special weld repair procedure may be required.	
Occasional minor weld defects.		Can be treated with existing qualified repair procedure.
Pipework		
Flowmeter lacks specified straight pipe length upstream	Engineering review needed. Do we reposition flowmeter or grant a concession?	
Pipe supports missing.		This is incomplete work.
Painting		
Significant deterioration over a large surface area.	Investigation needed to find out cause. Supervision policy to be reviewed.	
Minor paint damage in a number of places.		Can be repaired using existing repair procedure.

by whom they were solved. On the other hand, with a formal system in place:

- Nonconformities are identified as such.
- They are processed in a logical sequence by the appropriate persons through to completion.
- Recommendations/decisions/instructions/confirmation of close-out are in writing and duly signed all on a single form.
- Copies of all relevant supporting documents are included in the nonconformity file so that the record of each nonconformity is complete and self-contained.
- Recording all nonconformities in the one system gives an overview enabling tendencies to be detected revealing root causes previously unsuspected.

The best way to master nonconformities on site is for the EPCM contractor's quality assurance (QA) department to establish and run a site wide system covering the activities of all parties. Any of the latter, that is, the owner, the EPCM contractor or a construction contractor, can initiate a nonconformity. However, it is important that the nonconformity management system is closely supervised to prevent it from being flooded with trivial matters or contractual point scoring, which will tend to devalue it. Attention must also be paid to the introduction of the system, for the notion of nonconformity is often misunderstood, and not only at the lower echelons of an organization. A formal system can be perceived by some senior managers as highlighting problems and thus being detrimental to their public image. The response to this concern is that at least the system gives the means of clearly identifying anomalies and makes sure that they are appropriately processed. Corrective and preventive action will tend to root out underlying causes and thus improve effectiveness. An explanatory workshop given by the EPCM contractor site QA manager (SQAM), in the presence and with the full support of the site director, is a good way to evacuate any misgivings.

The resolution and close-out of nonconformities needs rigorous management; otherwise they can accumulate to become a significant cause of project delay. The EPCM contractor SQAM or one of his or her assistants must expedite the needed actions, especially when several parties are involved. This subject must be actively reviewed at every regular meeting between the owner and the EPCM contractor and between the latter and the construction contractors. Above all, the nonconformity management system must be perceived by top management of all parties as a vital tool for them to identify quality problems as they arrive and to follow them through to an acceptable documented outcome.

8.2 Resolution

A nonconformity may be resolved by choosing one or other of the following options:

- Reject, scrap and replace to specified requirements.
- Rework to meet completely specified requirements.
- Accept by granting a concession after repair, but still not fully meeting specified requirements.
- Accept as it is by concession with no rework or repair.
- Re-grade materiel for use in less arduous conditions.

This is illustrated in flowchart form in Figure 8.1, which shows the paths taken through to close out according to the model procedure set out in Section 8.3.

The above is written in terms applicable to materiel or works. For nonconformities centred on documentary or administrative matters, the same general principles apply. Resolution implies applying corrective action to the immediate problem, but also backtracking to find and correct the underlying cause. Preventive action can then be taken to preclude repetition of the same problem.

8.3 Model procedure

A model procedure is proposed hereunder based on the use of two simple forms:

- Nonconformity report (NCR), a model of which is shown in Figure 8.2

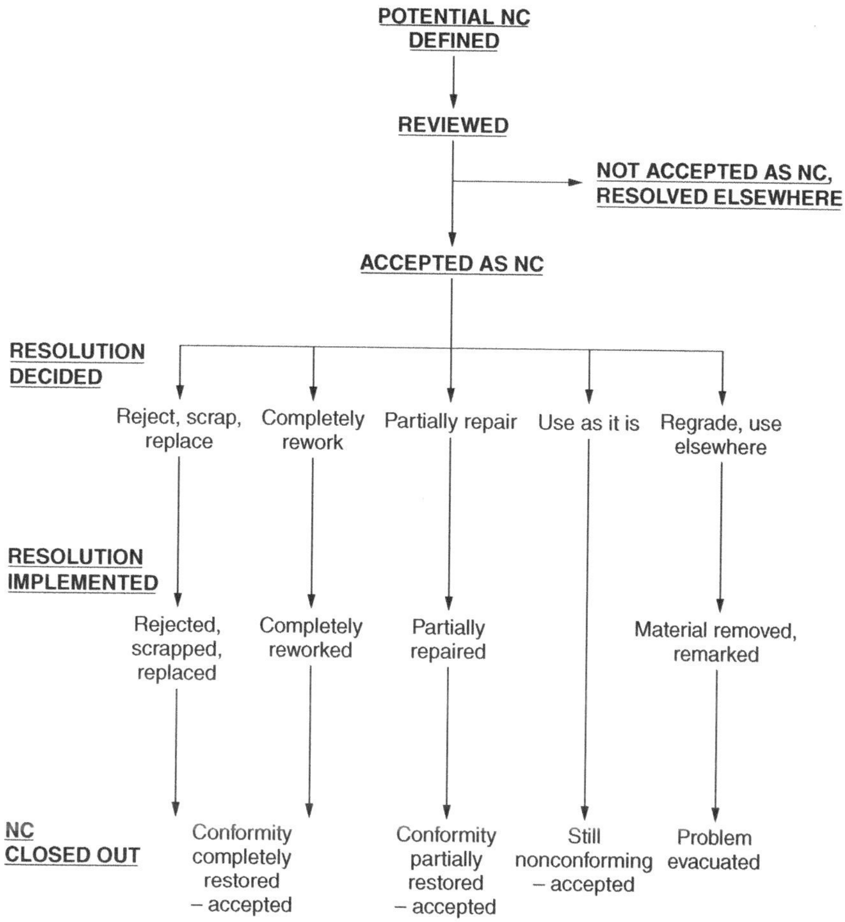

Figure 8.1 Nonconformity (NC) resolution and close out flowchart.

- Nonconformity log and status report (NCLSR), a model of which is shown in Figure 8.3

Each nonconformity is processed on an individual NCR, which records on a single sheet the successive stages of the processing of the nonconformity and carries the names and signatures of the persons involved. The information given on the NCR may be supplemented and validated by documents attached to and cross referenced to it. As each NCR is established it shall be entered on the NCLSR. The latter shall be continuously updated to reflect the status of all NCRs. It forms the basis for

EPCM contractor Name/Logo	**NONCONFORMITY REPORT** **NCR no. ....**	Owner: .... Project no.................. Project ....

A. DEFINITION OF NONCONFORMITY
Contract no. .... Material reference: ....
Reference document: ....

Originator
Representing: ....
Name: ....
Signature: ....
Appendices A1 to A .... attached
Date: ....

B. DEFINITION OF RESOLUTION
The following organization/department/person is requested to define the resolution of this nonconformity: ....

QA Department
Name: ....
Signature: ....
Resolution proposed:
Date: ....

Representing: ....
Name: ....
Signature: ....
Appendices B1 to B .... attached
Date: ....

C. RESOLUTION APPROVAL AND INSTRUCTION TO IMPLEMENT
The above defined resolution is approved. It shall be implemented by: ....

QA Department
Name: ....
Signature: ....
Appendices C1 to C............. attached
Date: ....

D. CLOSE-OUT
The resolution has been implemented.

Representing: ....
Name: ....
Signature: ....
Date: ....

QA Department
Name: ....
Signature: ....
Date: ....

Appendices D1 to D............. attached

Distribution:

Figure 8.2 Nonconformity report.

EPCM contractor Name/Logo	NONCONFORMITY LOG AND STATUS REPORT Updated: …../…../…..				Page: ……/……	Owner: ………………… Project no. …………… Project: …………………	
NCR no.	Contractor or dept. concerned	Contract no.	Originator	Definition of nonconformity	Resolution retained	Date established	Date closed out
Distribution:							

Figure 8.3 Nonconformity log and status report.

expediting NCRs and for reporting to management. A regular review of the NCLSR may reveal the same type of problem appearing repeatedly, perhaps concerning the same construction contractor, zone, discipline or activity. An investigation can then be instigated (by carrying out an audit for example) to identify and correct the underlying cause.

NCRs shall be numbered consecutively NCR 001, NCR 002, and so on. If considered useful, a suffix can be added to identify the geographical or process area/discipline concerned. This can facilitate sorting and classification of the NCRs.

We are now going to walk through the procedure for identifying and processing a nonconformity as follows:

- The originator completes and signs Section A of the NCR, attaching any relevant back-up documents clearly marked A1, A2, A3, and so on. *The problem, potentially a nonconformity, is thus defined.*
- The EPCM contractor's quality assurance (QA) department decides whether or not this can be classified as a nonconformity. If the decision is in the affirmative, it is entered in the NCLSR and a serial number is assigned to the NCR. *The nonconformity now exists.*
- Should the problem not be classed as a nonconformity, the QA department shall suggest to the originator an alternative for its resolution. *The problem is to be resolved outside of the nonconformity management system.*
- In Section B, the QA department designates the person to suggest an appropriate resolution.
- This person completes Section B proposing a resolution, if necessary attaching supporting documents marked B1, B2, B3, and so on.
- In Section C the QA department validates the proposed resolution and designates a person to implement it, backed-up as necessary with documents marked C1, C2, C3, and so on. *At this point the NCR defines what has to be done and who is responsible for seeing it is done, or in the case of a concession, indicates that nothing has to be done.*
- A copy of the NCR is sent to the designated person as an instruction to implement the corrective action.
- Section D records close out of the NCR, with back-up documents D1, D2, D3, and so on, if required, signed by the

person designated in Section C and counter-signed by the QA department.
- At each stage the NCLSR is updated to reflect the current status of the NCR.

The regularly updated NCLSR covering all nonconformities can be a tool for expediting and reporting to top management.

8.4 Few or many nonconformity reports

Suppose that on a process plant construction site, few NCRs are being registered, does it mean that this is a near perfect project running like clockwork, or should we suspect that the problems are there anyway, but are being resolved informally (in a nonrigorous fashion?), and perhaps even being covered up? Suppose that on another site many NCRs are being registered, can we suppose that this is a problem site, or should we congratulate the site management on identifying and resolving problems in a structured manner as they arise? Deciding whether there are too few or too many NCRs is subjective, so there is no point in citing figures from actual projects. However, if the system discovers individual problems and resolves them, the problems were there anyway; so, let us be happy that their discovery and resolution are properly recorded within a logical framework. If in addition analysis of the ensemble of nonconformities alerts management to a widespread, fundamental problem, this is even more beneficial.

CASE STUDY

Let us compare the experience of applying a formal nonconformity (NC) management system on two different process plant construction sites, A and B. On site A, from the day the site was opened, the NC management system was operational, an initiative totally supported by the EPCM contractor's site director, who considered it essential for mastering any quality problems that might arise.

Continued

CASE STUDY (Continued)

The system created was simple; processing a NC was done on a one-page form requiring a minimum of signatures. With full support from the top, no problems were encountered in imposing the system site-wide and it proved its worth throughout the whole construction phase.

On site B, although the EPCM contractor was certified to ISO 9001:2000, a status proclaimed by a framed certificate hanging on the office wall, and had a quality assurance manager on site, no system of NC management had been put in place. With the civil works finished and the erection of mechanical equipment well advanced, the owner became concerned and tried to impose such a system on the EPCM contractor. The NC report form created had to be of two pages to accommodate a cascade of signatures, which reflected the heavy decision-taking process imposed as part of the implication of the owner in site management detail. Even worse, key management figures on both the owner and EPCM manager sides, that is to say, the ones who ought to have been the driving force behind the new NC management system, were very sceptical and never understood what was a NC. Only a handful of NC reports were completed and some months later the scheme was abandoned. Quality problems continued to be treated, but not under the integrating umbrella of a formal system.

Lessons learnt

- Site A shows that when a NC system is simple, receives full support from the top and is operated competently, it is a very useful management tool.
- Site B is an example of a site with a complicated management structure, with no effective formal quality culture, and in which there was strong opposition to the introduction of a NC management system. Not surprisingly, it never worked. One reaction might be, 'So what! Does it matter?' Only a detailed investigation and analysis would determine the gains and losses, tangible and intangible, to both approaches. However, if by use of a simple system we can in an obvious transparent manner follow-through, determine and record the resolution of quality problems (nonconformities), can there really be any real reason to oppose it?
- As a final point, a framed ISO 9001 certificate hanging on the wall can mean a lot, but it can also, as in this case, mean nothing.

9

Quality Audits

9.1 Introduction

A quality audit is a systematic review of an activity, of works or of a procedure to compare the actual situation with what it should be to project requirements. This enables any anomalies to be identified and corrective action to be defined and implemented. It is also an opportunity to gauge the effectiveness of specified arrangements and to propose possible improvements. The audit should be carried out by a competent and experienced auditor in close collaboration with the individual(s) responsible for the matter being audited. Total support of both head office and site management is essential.

In Table 3.1 we have already indicated the various audits which could take place. From the perspective of the EPCM contractor on site, the main auditing thrust will be an ongoing programme of auditing the various construction contractors present. The EPCM contractor's site organization should also be audited, but this could best be done by its head office.

9.2 Guidelines for auditing, including a model procedure

Auditing is a sensitive area, which if not handled properly, generates resentment and noncooperation. Problems arise when the auditor displays a nit-picking attitude dramatizing minor irregularities, which damages his or her credibility and brings the whole idea of auditing into disrepute. So it is important that audits are conducted by persons of sufficient personal calibre and possessing the necessary people skills, in particular being able to communicate in a persuasive and diplomatic manner. A sense of humour is also an asset for gaining the confidence of the auditee. The full visible support of the EPCM contractor's site director is essential, which can best be demonstrated during a workshop held on site to present and explain quality auditing positively. An audit is not a blaming exercise, but rather a review to enable any problem areas to be detected and corrected for the benefit of all concerned, in particular for the auditee.

Figure 9.1 is a flowchart giving an overview of the quality auditing process. Below are set out some guidelines incorporating a procedure using the various model forms proposed in the figures attached to this chapter:

- Quality auditing shall be conducted in a nonconfrontational manner.
- Auditees on a construction site are busy people; so keep the audit short and to the point, preferably not exceeding half a day's duration as it is difficult to retain interest beyond this limit. Where a half day is insufficient, carry out two or more audits on different occasions.
- The EPCM contractor QA manager shall prepare and update a quality audit schedule. Figure 9.2 proposes a suitable format. This facilitates management of the over all audit programme.
- For a given activity typically three routine audits can be carried out, a preliminary one once the auditee has had time to set up his or her activity to project requirements, an intermediate one halfway through the activity and the final audit towards completion of construction.
- A special audit may be initiated in response to a situation giving rise for concern, for example loss of traceability in a piping prefabrication shop.
- The auditor shall agree with the auditee a mutually convenient date, time and place and shall confirm the

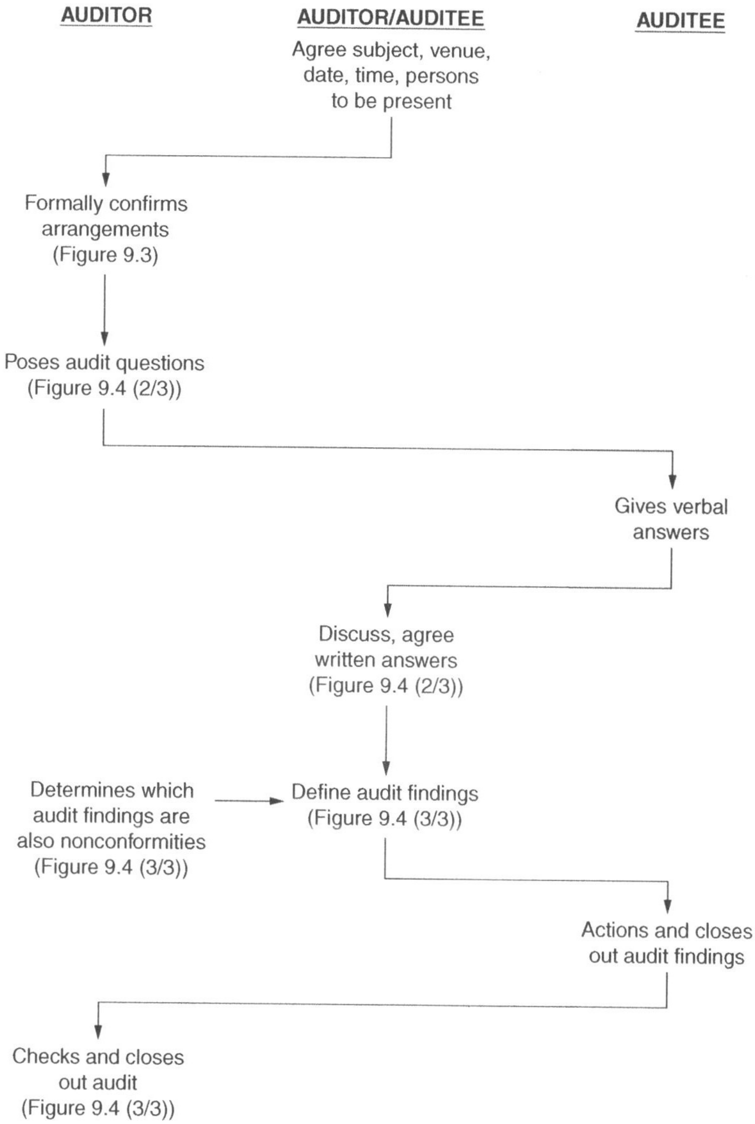

Figure 9.1 Quality audit flowchart.

EPCM contractor Name/Logo	QUALITY AUDIT SCHEDULE Updated: ...../...../.....		Page: ....../......															Owner: .... Project no. .... Project: ....
Organization to be audited	Contract no. (where applicable)	Subject and place of audit	Audit scheduled for week beginning Monday:															Actual date of audit
			*															
			**															
Distribution:	*Enter date **Enter 'P' for 'Preliminary', 'I' for 'Intermediate', 'F' for 'Final' & 'S' for 'Special'																	

Figure 9.2 Quality audit schedule.

arrangement in writing. Figure 9.3 proposes a suitable notice form. At this point the auditee, in view of the forthcoming audit, may well review his/her activity and correct any irregularities before the audit takes place. If this occurs, so much the better.

- For the sake of credibility and relevance, an audit must be centred on concrete issues defined as a written requirement. By the same token, audit questions must be framed in such a manner as to lead to factual objective answers.
- Before any questions are asked, a tour of the site where the activity takes place shall be made and a return to the site shall be made as and when required during the audit for any clarification.
- The three-page Figure 9.4 proposes model forms for the front page, the one or more questionnaire pages and the conclusion page of the audit report.
- The auditor shall write his questions on the left hand side of the audit form Figure 9.4 (2/3), but shall only reveal them to the auditee during the audit.
- The subject matter of each audit shall be jointly investigated by the auditor and the auditee. An agreed response to each question shall be entered on the right-hand side of the form. If the question is properly framed and the auditor is experienced, there should be no difficulty in agreeing a response.
- Should the audit reveal a shortcoming on the part of the auditor's own organization, it shall be duly noted and actioned in a timely manner.
- At the end of the audit both parties shall review the completed questionnaire to identify anomalies to be designated as audit findings (AFs), numbered AF1, AF2, and so on. The more significant AFs may, where appropriate, be classified as nonconformities to be entered into the nonconformity management system described in Chapter 8.
- Each AF shall be entered on the conclusion sheet (Figure 9.4 (3/3)) together with a definition of corrective action and a target close-out date. Both the auditor and the auditee shall sign the conclusion sheet.
- The audit report shall be completed, signed and distributed within two working days of the completion of the audit, any delay lessening the impact.

EPCM contractor Name/Logo	**QUALITY AUDIT NOTICE**	Owner: .... Project no. .... Project: ....

AUDIT No. ....

Name of organization to be audited: ....

Contract/P.O. no.: ....

Title of contract/P.O.: ....

Location/address where audit is to be executed: ....

....

Date/time of audit: ...../...../............. at .... hours

Subject of audit: ....

Reference documents to be used as basis of audit:

....

....

....

We confirm our intention to audit your organization as defined above and we look forward to receiving your fully cooperation in this matter. Would you please arrange for the following members of your staff to attend the audit:

....

....

The auditor(s) will be:

....

For EPCM contractor

Name:..

Signature: ....

Date: ....

Distribution:

Figure 9.3 Quality audit notice.

EPCM contractor Name/Logo	**QUALITY AUDIT REPORT** **(Page 1/....)**	Owner: .... Project no. .... Project: ....

AUDIT No. ....

Name of organization audited: ....

Contract/P.O. no.: ....

Title of contract/P.O.: ....

Location/address where audit was executed: ....

....

Date of audit: ....../....../..........

Subject of audit: ....

Reference documents forming basis of audit:

....

....

....

Participants:

Name	Organization	Function
....	....	....
....	....	....
....	....	....
....	....	....
....	....	....

Distribution: 1/3

Figure 9.4 Quality audit report in three parts (1/3, 2/3 and 3/3).

EPCM contractor Name/Logo	**QUALITY AUDIT REPORT** (Audit no. .... Page ..../....)	Owner: .... Project no. .... Project: ....

QUESTION/POINT RAISED	**AGREED ANSWER**
	2/3

Figure 9.4 (Continued)

EPCM contractor Name/Logo	**QUALITY AUDIT REPORT CONCLUSION** (Audit no. .... Page: ....../......)	Owner: .... Project no. .... Project: ....

Following a joint review of the completed questionnaire in this report, it was agreed that the undermentioned constitute the audit findings (AFs) requiring corrective action to close out. AFs of a more serious nature may be classed as nonconformities. These are assigned a NCR no. and are entered in the site nonconformity management system.

AF no. (NCR no.)	Audit finding	Corrective action	Target date	Close out date	Initials

.... pages attached as appendices marked up with AF no. as appropriate

	Organization	**Represented by**	**Signature**	**Date**
Auditor:	....	....	....	...../...../......
Auditee:	....	....	....	...../...../......

3/3

All AFs closed out
Name: ....
Signature: ....
Date: ....

Figure 9.4 (Continued)

- Distribution of the audit report shall be restricted to those needing to know; this helps the auditee to accept more readily the fact of his or her shortcomings being brought to light.
- The audit report lists problems to be corrected, but the cover note for the distribution can very well convey an over all impression giving praise where praise is due.
- An audit is useless if it is not expedited to close out as early as is possible. The conclusion sheet (Figure 9.4 (3/3) includes two columns to record the close-out of each AF.
- Figure 9.5 is a model audit log and status report to register the audits and record progress to close-out.

9.3 Typical audit questions

Since an audit is to be based on specified requirements, before preparing his/her questions the auditor must study the relevant contract, procedures, method statements, quality plans, specifications, standards and any other pertinent documents. However, for general guidance typical audit questions are given hereunder:

Quality records (see Sections 5.5, 5.7 and 5.8):

- Are inspection and test plans covering the full work scope approved and available?
- Do inspection and test record forms covering the whole workscope exist, and have they been partially completed for each item of equipment in anticipation of the specified inspections and tests to be carried out?
- Is the format and content of the construction quality file (CQF) defined?
- Has the CQF been created and does it receive completed and signed off forms as the works progress?
- Who actually fills in the forms initially, by whom are they signed; are they signed off within 24 hours of the completion of actual inspections and tests?
- For prefabricated items, what is the procedure for ensuring that all shop inspection and testing have been satisfactorily completed and recorded before items are shipped to site?

<table>
<tr><td colspan="2">EPCM contractor

Name/Logo</td><td colspan="4"><u>QUALITY AUDIT LOG AND STATUS REPORT</u>
Updated: ....../....../...... Page: ...../.....</td><td colspan="2">Owner: ....
Project no. ....
Project: ....</td></tr>
<tr><td rowspan="2"><u>Audit no.</u></td><td rowspan="2"><u>Organization audited</u></td><td rowspan="2"><u>Date of audit</u></td><td colspan="2">No. of audit findings</td><td rowspan="2"><u>Date closed out</u></td><td rowspan="2" colspan="2"><u>Remarks</u></td></tr>
<tr><td><u>Initially</u></td><td><u>Outstanding</u></td></tr>
<tr><td></td><td></td><td></td><td></td><td></td><td></td><td colspan="2"></td></tr>
<tr><td colspan="8"><u>Distribution:</u></td></tr>
</table>

Figure 9.5 Quality audit log and status report.

- How are quality records from special service providers integrated in to the quality dossier?
- Are the quality record requirements for authorities approvals clearly defined?
- Are the quality records necessary for contractual 'construction completion' and 'turnover to the owner' clearly defined?
- Are records kept in a secure protected environment?
- Is there a second copy in a second location?
- What arrangements are made for safe storage of photographic film?
- What arrangements are made for safeguarding electronic records?

Materiel control (see Sections 7.3–7.5, and Figures 7.1–7.4):

- Is the relevant materiel control procedure, duly approved, available in the warehouse office?
- Are the materiel control posts shown on the organization chart duly filled?
- Are the computer hardware and software installed, and have personnel received appropriate training?
- Are the warehouse and yard enclosed and fully equipped?
- Are spaces in the yard and in the warehouse identified with a grid reference to define materiel location?
- How is incoming materiel received and processed?
- What are arrangements for incoming technical inspection?
- Is there a quarantine area for nonconforming materiel?
- In the case where defective materiel arrives on the site, how is it replaced or repaired?
- Do certificates arrive with the materiel, and if so how are they processed and how is traceability maintained?
- How is materiel issued?
- Is there a formal procedure for ongoing inspection/preservation of certain items of equipment (vessels under inert gas purge or electric motors requiring regular shaft rotation for example)?
- Is this procedure in fact implemented?

Reinforced concrete works (see Appendix A):

- Who is responsible for supervision of the works and for quality control?
- Do the number and qualifications of these individuals correspond to the site quality plan?
- Are the relevant drawings, specifications, standards and codes of practice issued to these persons, or are they at least readily available for reference?
- Is there an approved batching plant producing concrete to the approved mix design(s) using approved sources of material?
- Are the details and references of the approved mix design(s) displayed in the batching plant?
- Are the reinforcing bars subjected to incoming inspection including checking of steel grade through traceability to mill certificates?
- What pre-pour inspection is carried out, by whom and how is it recorded?
- By whom is the pouring supervised and witnessed?
- How is curing supervised?
- Who is responsible for inspection after removal of formwork and how is it recorded?
- How are concrete truck deliveries supervised for timing, quality (slump test) and traceability to the delivery ticket?
- Who takes concrete cylinder/cube samples, at what frequency, and how are they marked, cured and delivered to the laboratory?
- Which laboratory carries out crushing tests on these samples, and how are the test results recorded?
- What is the traceability trail between raw materials, mix design, batch plant, truck and load numbers, the concrete foundation or structure in question, the samples and the test results? (See Figure A1).
- Who is responsible for reviewing and accepting the laboratory test results?
- What arrangements are made to review trends as opposed to merely checking a series of individual test results?

CASE STUDY

Over a period of years some 200 quality audits were carried out on several process plant construction sites. About 10 of these were directed within the EPCM contractor's organization, the remainder concerning construction contractors. They give rise to the following comments:

- While there was on occasions some reticence initially on the part of the auditee, there was never a problem at the end of the audit to agree the text of the audit report, in particular the formulation of the audit findings, that is to say, anomalies needing correction, because the points noted were concrete ones, tied back to contractual documents.
- About 10% of the audits revealed only minor problems or none at all, another 10%, quite serious problems, and the 80% majority brought to light a range of anomalies of varying significance, which needed correction.
- Taking one particular case, on an electrolytic refining project, quality problems were regularly coming to light concerning certain materials used for busbar prefabrication on site. An in depth audit conducted at short notice by the EPCM contractor on the manufacturer's premises revealed no process irregularities. In addition, the organization of quality management was sufficient and appropriate for the type of manufacturing operation concerned. However, from the time of the audit onwards, the problems never reappeared. Clearly, under the pressure of the forthcoming audit, the manufacturer had put its house in order. We shall never know what had been wrong and what had been done to correct it, but nevertheless the audit produced a satisfactory outcome.
- Auditing made most impact where there was top management support, leading to timely correction of audit findings.

Lesson learnt

Auditing can make a very useful, even essential and vital, contribution to quality management on a process plant construction site, but it has to be properly conducted and receive top management support from all parties involved. This can lead to auditing being accepted as something positive, and not just as a police operation.

10

Management Reviews and Completion Report

10.1 Introduction

The site quality management system must be kept under surveillance by the EPCM contractor's site quality assurance manager to identify areas where modification or development can bring about an improvement in system performance. This will be an ongoing everyday process, but it is desirable to establish a number of milestones for the EPCM staff to review operations on site in a formal manner. This can happen at two levels:

- Periodic management reviews aimed at improving the situation for the project in hand
- As construction draws to a close and before staff start to be demobilized, the preparation of a construction completion report, so that future projects can benefit from lessons learnt

Whilst this chapter is focussed on the EPCM contractor, the same principles apply to other parties present on site.

10.2 Management reviews on site during construction

During the construction period, management reviews should be scheduled at regular intervals. It is the task of the site quality assurance manager (SQAM) to organize these reviews. Inputs should be sought from all members of the management team directly involved, and through them opinions can be obtained from subordinate staff. Where appropriate the owner, construction contractors, suppliers' representatives and special-service providers may also be asked for comments on specific issues. If the review is to be constructive and comprehensive, it has to be managed and animated by the SQAM so that conclusions can be presented as precise points for action, with definition of responsibilities for expediting through to close out. One way to set this in motion would be to use the form shown in Figure 10.1. The SQAM can enter the relevant questions and discussion points in column 2 and distribute it to all parties, asking each manager to address the questions related to his or her responsibilities, and draft answers in column 3. A meeting (more than one if necessary) can be held to determine and record details of action required, the name of the person responsible and a target date for completion. Where appropriate, 'no action required' can be entered against an item. The definitive review form is distributed to all participants and the SQAM expedites them to close out.

The details will of course correspond to the situation on the site in question, but looking through this handbook we can generate typical questions and suggest areas for review, for example:

- Are the actions decided during the last management review completed and closed out?
- Interfaces with the owner (reports, meetings, regular joint site visits, informal contacts).
- The EPCM contractor's organization chart, distribution of responsibilities.
- The EPCM contractor's human and material resources.
- Are site visits of suppliers' representatives controlled and recorded?

EPCM contractor Name/Logo	**QUALITY MANAGEMENT SYSTEM REVIEW REPORT** Date: ..../..../...... Page: ..../.....			Owner: .... Project no. .... Project:	
Item no.	Question raised	Action decided	Person responsible	Target date	Close out date
Distribution:					

Figure 10.1 Quality management system review report.

- Are surveying, laboratory and inspection services on the site satisfactory?
- Have satisfactory relationships been established with the utility companies?
- Similarly are there satisfactory relationships with the authorities?
- Also with the insurer?
- Has each construction contractor a site quality plan, an inspection and test plan, procedures and method statements, model inspection and test record forms and the definition of the contents of the construction quality file?
- Is each construction contractor keeping records of inspections and tests and filing them in the construction quality file?
- Is there a procedure for regularly calibrating inspection, measuring and test equipment, and is it effectively applied?
- Is document control functioning properly, including interfacing with head office?
- Breakdown of site-generated engineering queries: what does it tell us?
- Has the policy for 'as-built' drawings been established? Has it been put into effect? How is it supervised?
- Is the site element of the project information technology infrastructure in place and operational?
- Is the EPCM contractor's materiel control system in place and operational: warehouse and yard, staffing, computers and telecommunications, procedures?
- Are materiel incoming inspection, ongoing inspection and maintenance being carried out and recorded to an approved procedure?
- Have the construction contractors set up appropriate materiel control systems for the materiel they procure?
- Are the requirements for traceability defined? Who is responsible for ensuring they are carried out?
- A complete breakdown of nonconformities by discipline and per each organization. What underlying trends can be detected?
- A similar breakdown for audit findings and a similar question to be asked.

- Are there agreed-upon definitions for 'construction completion', 'turnover', 'commissioning', 'start-up', 'performance testing' and 'commercial operation'?
- Is the procedure for 'turnover to the owner' fully agreed?
- If turnover is to be made system by system, are all the systems fully defined?
- Are format and content of final documentation fully defined?
- What is the situation concerning spare parts for commissioning, start-up and commercial operation?
- Is the owner taking steps in terms of staffing, organization and material resources to be able in effect to take over the plant?
- Looking beyond turnover, what other services will the owner expect of the EPCM contractor and other parties present?

10.3 Construction completion report

As the construction phase draws to a close, steps must be taken to record what happened and the lessons to be learnt. This is in the longer-term interests of the EPCM contractor enabling future projects to benefit. Whilst much can be learnt from a systematic review of quality records, nonconformity reports, audit reports and so on, the most potent source of information will be the EPCM contractor's site team members. They must be debriefed before they are demobilized. The difficulty is that this has to be done at a time of maximum activity as outstanding items are being finished, final inspection and testing are reaching a peak and turnover to the owner is imminent. Where possible, a suitable person should be assigned to the task of preparing the report on a full-time basis, being relieved of other duties. One approach to debriefing the personnel would be to prepare a questionnaire based on much the same subject areas as cited in Section 10.2, asking the concerned persons for each subject:

- What went well?
- What did not go so well?

- Based on this experience what recommendations would you make for future projects?

After further dialogue and investigation, all of these sources of information can be collated and edited into a comprehensive report.

11

Construction Completion and Turnover

11.1 Activity phases on site

Activities on site will pass successively through the phases of construction, pre-commissioning, commissioning, start-up, performance testing and, finally, commercial operation. These terms can give rise to different interpretations from one organization to another; so, it is important to define them formally at the beginning of a project by listing the scope of work covered in each phase. Equally important is to define at which point and how responsibility for the site is to be transferred from the EPCM contractor to the owner. A brief definition of each phase is given hereunder, being an approach in common use.

Construction comprises all the activities needed to build the plant, including the erection of equipment, together with the inspections and tests carried out throughout this phase.

Pre-commissioning covers the static checks and preparation works carried out on the completed plant.

Commissioning is essentially all actions to prepare the completed plant for start-up and to prove the integrity of the plant without the introduction of raw materials.

Start-up is the initial attempts to run the plant for the purpose for which it is intended, that is, from an input of raw material to give an output of the finished product.

Performance testing is to prove the ability of the complete plant to produce within the parameters defined contractually.

Commercial operation, which is what it says, will be able to be achieved once the plant has been proved and any problems resolved.

11.2 Contractual milestones

Pre-commissioning activities can be readily grouped with construction properly speaking to be carried out by the construction contractors, who possess the necessary skills. These activities typically include:

- Flushing and blowing through piping
- Pressure testing, then reinstating piping
- Testing relief valves
- Installing internals in vessels before closing them
- Aligning driver/driven machines
- Lubrication and gland packing
- Load tests on lifting devices
- Checking electric motors for direction of rotation
- Checking electric circuits and adjusting protective devices
- Loop checks on instrumentation and control systems

When the scope of work of a construction contractor is finished, the contractual milestone of construction completion can be declared (see Section 11.4).

However, from commissioning onwards operations are of a different nature, being for the most part, outside the normal competence of construction personnel. Instead, there is need of a team of persons possessing a range of knowledge and skills spanning plant design, in particular the process, plant operations, plant maintenance, safety and, increasingly, information technology. This team can be in large part made up of the individuals who are going to run the plant in the future, backed up with specialists from a number of different sources. For this reason, on many projects, the contractual watershed milestone of turnover from the EPCM contractor to the owner takes place upon completion of construction (including pre-commissioning) and before commissioning. This is the arrangement assumed in this handbook.

Discussion of the subsequent phases of commissioning, start-up and performance testing leading to commercial operations is beyond the scope of this handbook. Should the EPCM contractor still be implicated, then there will be further contractual milestones to be established. However, even if there is no direct responsibility, the EPCM contractor and the construction contractors may be required to fulfil a supporting role as evoked in Section 11.3.

11.3 Responsibilities of parties present

If the situation on site is not tightly controlled, one phase will tend to merge into the other. However, it is absolutely essential that contractual milestones are defined, understood, executed and respected by all parties, in particular the watershed milestone which is turnover to the owner. From that moment onwards the owner assumes complete responsibility for all operations on site. Electric power and control circuits will be energized, machinery will be operational, and potentially dangerous fluids may be present. Consequently, the site occupational health and safety and environmental management procedures must be upgraded to take account of the new situation, in particular concerning 'permit to work' arrangements.

However the EPCM contractor's and the construction contractors' site teams may still be involved, if required by the owner, in a supporting role to:

- Complete any items outstanding
- Remove insulation to give access and reinstate it afterwards
- Provide teams to carry out modifications and additions as may be required

They will now work under the constraints imposed by the owner's management, in particular within a work permit/ safety system.

Depending on the rapidity or otherwise with which the owner is able to render fully operational its own on-site departments, the EPCM contractor may be requested to continue for the time being to provide assistance in some or all of the following domains:

- Resolving engineering queries
- Updating drawings to 'as-built' status (during commissioning and start-up there may well be a lot of modifications to be recorded)
- Liaison with suppliers for spare parts
- Liaison with suppliers for site visits of specialists
- Management of specialist services such as nondestructive examination
- Technical liaison with the authorities

11.4 Construction completion procedure

Let us first consider the case in which the procedure is applied to the whole plant at the same time (or to a distinctive independent self-contained unit within a larger plant). There will normally have been a number, perhaps a considerable number, of construction contractors involved. Verification of construction completion and its contractual formalization can be achieved applying the following procedure for each construction contractor:

- When a construction contractor considers that the works for which he or she is contractually responsible and the associated documentation are complete, the construction contractor shall invite the EPCM contractor to carry out an inspection using the model form presented in Figure 11.1.
- Representatives of the EPCM contractor and the construction contractor jointly inspect the works and review the corresponding documentation, any uncompleted works or documentation discovered being entered on a list of

EPCM contractor Name/Logo	**CONSTRUCTION COMPLETION** **REQUEST FOR INSPECTION**	Owner: .... Project no. .... Project: ....

Construction contractor: ....

Contract no. ....

Contract title: ....

This request concerns the following works/equipment/functional systems*:

....

as further described in the following documents copies attached (........... pages):

....

....

We the undersigned representing the construction contractor, consider that construction operations associated with the above defined works/equipment/functional system(s)* together with the corresponding quality records, have been completed according to the terms of the contract.

In consequence we request the EPCM contractor to inspect jointly with us the works/equipment/functionl system* and to review the associated quality records with a view of establishing a construction completion certificate.

For construction contractor

Name:.......................................

Signature: ....

Date: ....

*Delete as necessary.

Distribution:

Figure 11.1 Construction completion request for inspection.

outstanding items (LOI), a model of which is presented in Figure 11.2, duly signed by both parties.
- Any relevant unresolved nonconformities or audit findings (Chapters 8 and 9) can be simply transferred to the LOI.
- The EPCM contractor expedites the construction contractor until the LOI is closed out, or at least reduced to a few items not sufficiently significant to block the proceedings.
- When the EPCM contractor is satisfied, the construction completion certificate, a model of which is presented in Figure 11.3, is prepared and signed by the construction contractor and countersigned by the EPCM contractor.
- The construction contractor must continue to progress the work as far as it can on items still on the LOI, if there is one, and the EPCM contractor will expedite accordingly. The LOI's last two columns are intended for noting close out item by item.

The above procedure will be applied to all the construction contractors at more or less the same time, which means that following up the various LOIs may involve a lot of activity on the part of the staff of the EPCM contractor. So to keep a grip on the situation, the LOIs should be numbered and over all control maintained using the list of outstanding items log and status report as presented in Figure 11.4. (LOIs are also known as 'punch lists' and identifying outstanding items and entering them on a LOI is called 'punch listing'.)

When a construction completion certificate has been signed for every contract, construction can be deemed to be complete for the whole plant.

11.5 Procedure for turnover to the owner

The EPCM contractor should now be in a position to turnover the whole plant to the owner according to the following procedure:

- From the ensemble of the LOIs attached to the construction completion certificates received from the various

EPCM contractor Name/Logo	**LIST OF OUTSTANDING ITEMS** List no. .... Updated: ..../..../.... Page : ..../....		Contract no./Discipline/Works/System		Owner: .... Project no. .... Project: ....
Item no.	Location/Identification	Description	Target date	Close out date	Remarks
All items on this page located/identified/agreed. **Representing: ....** Name: .... Signature: .... Date:		**Representing: ....** Name: .... Signature: .... Date:		All items on this page closed out **Representing: ....** Name: .... Signature: .... Date:	
Distribution:					

Figure 11.2 List of outstanding items.

EPCM contractor Name/Logo	**CONSTRUCTION COMPLETION CERTIFICATE**	Owner: .... Project no. .... Project: ....

Certificate no.: ....

Construction contractor: ....

Contract no. ....

Contract title: ....

This certificate concerns the following works/equipment/functional system(s)*:

....

as further described in the following documents, copies attached (.......... pages):

....

....

We the undersigned construction contractor certify that for the above defined works/equipment/functional system(s)*:

A. All the works are complete
B. All inspection and testing have been completed
C. Operations A and B have been duly recorded on approved forms signed by both ourselves and the EPCM contractor and filed in the construction quality file
D. Modifications effected on site have been recorded on marked-up drawings to show 'as-built' status and filed in the construction quality file.

except as noted on the attached list of outstanding items No. .... of .... pages.

For construction contractor
Name:..
Signature: ....
Date: ....

We the undersigned after verification confirm that the above cited operations A, B, C and D have been completed, except for the attached list of outstanding items, which only includes minor points, or ones that for practical reasons cannot be completed for the time being. It nevertheless remains the responsibility of the construction contractor to complete them at the earliest opportunity.

For EPCM contractor
Name:..
Signature: ....
Date: ....
*Delete as necessary.
Distribution:

Figure 11.3 Construction completion certificate.

<table>
<tr><td colspan="2">EPCM contractor

Name/Logo</td><td colspan="3">LIST OF OUTSTANDING ITEMS
LOG AND STATUS REPORT
Updated: ....../....../...... Page: ...../.....</td><td colspan="3">Owner: ....
Project no. ....
Project:.........................</td></tr>
<tr><td rowspan="2">List no.</td><td rowspan="2">Contract no.</td><td rowspan="2">Date established</td><td colspan="2">No. of items</td><td rowspan="2">Date closed out</td><td rowspan="2" colspan="2">Remarks</td></tr>
<tr><td>Initially</td><td>Outstanding</td></tr>
<tr><td></td><td></td><td></td><td></td><td></td><td></td><td colspan="2"></td></tr>
<tr><td colspan="8">Distribution:</td></tr>
</table>

Figure 11.4 List of outstanding items log and status report.

construction contractors, the EPCM contractor produces a single consolidated LOI.

- The EPCM contractor shall continue to expedite all items, until the consolidated LOI is closed out or reduced to a minimum acceptable level.
- The EPCM contractor invites the owner to inspect the whole of the works and to review all the construction completion certificates and their associated documentation.
- The EPCM contractor completes a turnover certificate as per the model proposed in Figure 11.5 and both parties sign it.
- The whole plant has now been turned over to the owner.

If the representatives of the owner have been in sufficient strength as observers on the site during the completion of the construction phase as outlined in Section 11.4 above, and are thus familiar with the detail of the proceedings, the turnover operation should be straight forward. The procedures for construction completion and turnover to the owner are shown as a flowchart in Figure 11.6.

11.6 Construction completion and turnover by functional systems

The procedures to achieve construction completion and turnover the whole plant to the owner as a single operation as defined in Sections 11.4 and 11.5 can work for a smaller or not particularly sophisticated plant, with not too tight a schedule. However, the complexity and size of the plant may be such that a choice is made to organize construction completion and turnover to the owner spread out over a period of time on the basis of functional system by functional system. This enables utilities to be made available early and other systems to be processed in order of priority determined by the needs of start-up and commissioning. However, there is a difficult transition that we must negotiate, for the logic of division by functional system, which is at the heart of commissioning and start up, cuts across the logic of construction. That

EPCM contractor Name/Logo	**TURNOVER CERTIFICATE**	Owner: .... Project no. .... Project: ....

Certificate no. ....

This certificate concerns the following works/equipment/functional system(s)*:

....

....

completed by the following construction contractors:

Contract no.	**Construction contractor**	**Construction completion certificate no.**
....	....	....
....	....	....
....	....	....
....	....	....
....	....	....
....	....	....
....	....	....

We the undersigned, representing the EPCM contractor, and representing the owner, having jointly inspected the above defined works/equipment/functional systems*, and having reviewed the above listed construction completion certificates, including the supporting documentation, and the attached consolidated list of outstanding items, copies all of which are attached, declare the works to be transferred this day to the owner. In consequence care, custody and control of the works, from now on are the responsibility of the owner, in particular for matters of occupational health and safety and impact on the environment

For EPCM contractor	**For owner**
Name:	Name: ....
Signature:	Signature: ....
Date:	Date: ....

*Delete as required.

Distribution:

Figure 11.5 Turnover certificate.

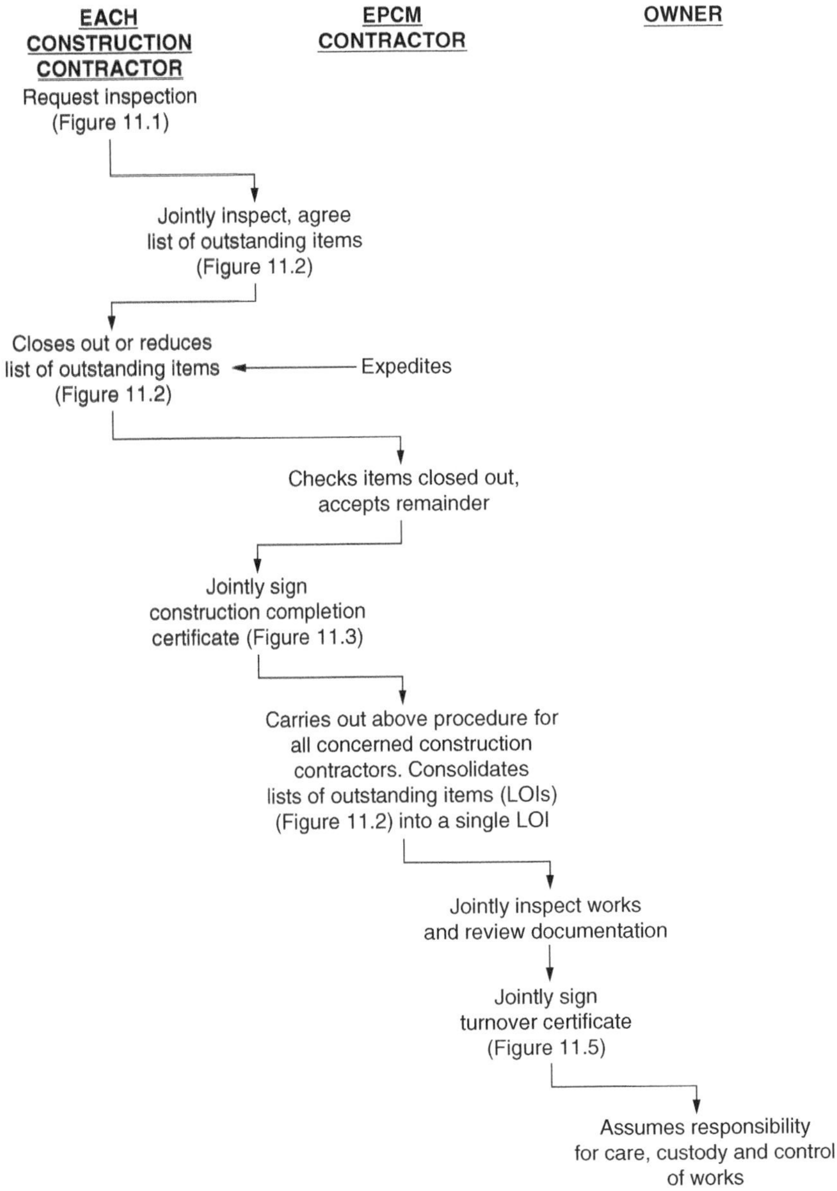

Note: The construction completion and turnover procedure illustrated in this flowchart can be applied to the whole plant, or to a part of it, in particular to a functional system.

Figure 11.6 Construction completion and turnover flowchart.

is to say that each construction contractor will be involved in a number of, possibly many, functional systems and each function system will comprise works carried out by a number of different construction contractors. This notion is illustrated in Figure 11.7, 'contracts versus systems'. Unless obliged to do otherwise, a construction contractor will arrange his operations to optimize construction, which probably would not be consistent with the order of priorities for completion by functional systems. So how do we solve this dilemma?

There is a simple and obvious answer to this question, and that is to embed the definition of functional systems as early as possible into the design process, with each system being allocated an alphanumeric code, which can be integrated into the overall project identification coding system. The components in each discipline, which are part of each functional system, must be identified. The definition of each functional system and the required order of completion of construction of functional systems will be introduced into the planning process and into the construction contracts from the call for bids stage onwards. Thus, each potential, and then each appointed construction contractor, will have made allowances for this and plan his inputs into the contract accordingly. Thus, introducing the notion of functional systems into the project at an early stage will contribute to ensuring a smooth turnover and facilitate the subsequent start-up and commissioning.

However, the timely addressing of this question does not always happen. At the beginning of the project, construction completion and turnover are a long way off, and the decision to complete and turn over functional system by functional system may only be taken at a later stage when construction is well advanced. Depending on the type of plant, the materiel content of each functional system, with a definition of its boundaries and interfaces with other systems, will have to be derived from a review of documents such as:

- Process and instrumentation diagrams (P&IDs)
- Single-line diagrams
- List of piping isometric drawings
- Pipeline lists

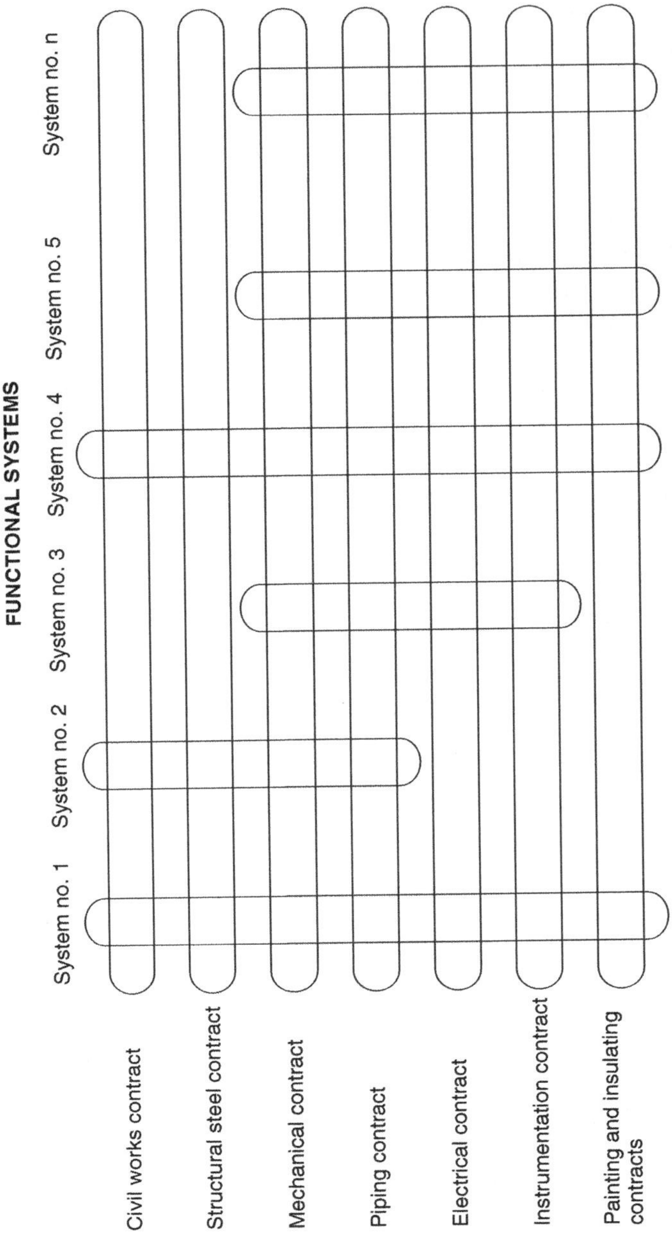

Figure 11.7 Contracts versus systems.

- List of vessels and tanks
- List of mechanical equipment
- List of electric motors
- List of switchgear and motor controls
- List of cables
- List of instruments
- List of instrumentation and control loop diagrams

Information can be extracted from these to generate specific new documents, or it may sometimes, in the case of a small project, be simpler to annotate existing documents. The P&IDs can be marked up to define functional system limits. If this approach is decided, it is very important to recognize from the outset the very considerable amount of time and effort required.

A typical example of a functional system to be so processed is a river extraction cooling water system. The construction contractors for the various disciplines will be responsible for completion of the following works:

• Civil contract	Intake, sump and pump support structure, discharge
• Structural steelwork contract	Crane support structure
• Mechanical contract	Pump sets, intake screens and filters, valves, manifold, crane
• Piping contract	Piping prefabrication and erection with supports, connections to heat exchangers (interfaces with other systems)
• Electrical contract	Cabling and connections, earthing, cathodic protection*
• Control and instrumentation contract	Low suction level and vibration protection on pumpsets, pressure and flow measuring systems, cabling, etc.
• Painting contract	Finish painting on equipment, blast and paint on structural steelwork and pipework

The asterisk means that the corresponding motor starters could be included in this system or alternatively in one of the electrical systems with the associated substation.

For the functional system-by-system case, the flow diagram in Figure 11.6, in principle, still applies, except that it does not apply to the whole plant, but is repeated for each system.

11.7 Conclusions

The reaction of the reader might well be that the above proposed model procedures for construction completion and turnover to the owner, particularly those based on functional systems, are complicated, but whatever the approach, managing completion and turnover is never simple. However, these procedures can work well provided that:

- The decision to operate system by system is taken early in the project in complete agreement between the EPCM contractor and the owner.
- The content and boundaries of each system are determined early and are integrated into project engineering and documentation.
- The order of priority of construction and turnover of the systems is defined.
- Construction contractors know from the bid stage onwards that construction completion will be achieved and documented system by system, and that they will be encouraged to proceed expeditiously along this path by linking substantial stage payments to this exercise.
- A single competent interlocutor, if necessary on a full-time basis, is designated within the site organizations of the owner, of the EPCM contractor and of each construction contractor; they shall collaborate closely.
- The necessary resources shall be engaged to manage and expedite the outstanding items to eventual close out.
- Most important of all, top management of all parties shall be resolutely committed to a dynamic approach leading to

the timely turnover of the complete plant, which on the owner's side means having developed a team capable of accepting the turnover and taking it forward through commissioning and start-up to commercial operation.

CASE STUDY

On a process plant construction site, the erection of mechanical equipment was finished, piping was almost finished, and electrical installations and instrumentation and control systems were well advanced. Functional system content and limits had been defined only recently. The construction contractors were under no contractual obligation to complete their work prioritizing by systems. They considered that to do so would disturb their planning schedules and manning arrangements. A large number of minor items were incomplete and it was difficult to oblige the construction contractors concerned to attend to them, without becoming bogged down in protracted discussions about extra payment.

For the owner time was pressing, so it was decided to simplify the project structure. In the first place the owner brought in one of its own engineers to head up the operation. He already possessed the experience of several plant construction completions and turnovers, had developed a procedure and came armed with computer software to track the operation. An integrated team was formed, comprising staff from the owner and the EPCM contractor. The contracts with the construction contractors for the mechanical, piping, electrical and control and instrumentation disciplines were closed out as such, and instead these firms made available teams of pipe fitters, welders, electricians, etc., to be integrated into the owner–EPCM contractor team. Their services would be paid for on the basis of time spent at hourly rates.

A team was assigned to the inspection of each system covering all disciplines, in order of priority, so as to identify, locate and record unfinished works on lists of outstanding items (LOIs). These items were identified *in situ* by means of colour-coded labels and numbered to be traceable to the functional system and the LOI concerned. Reports could be printed out in several different formats, in particular enabling items to be sorted by discipline for work assignment to the

Continued

CASE STUDY (Continued)

different trade teams. Inspection and testing were carried out and the records thus generated were filed in the construction quality files already established and partial completed by the construction contractors. As functional systems were turned over to the owner, they were physically identified and access to them was forbidden by means of tapes, labels and notices. The safety and work permit arrangements were strictly enforced to take account of the possibility that in some places completion of construction and commissioning were proceeding simultaneously in close physical proximity. The programme of construction completion and turnover was completed without too many problems.

Lessons learnt

- Not having identified functional systems much earlier complicated the situation.
- Interrupting the contracts before the construction contractors had themselves declared their works complete gave rise to a massive task of seeking out uncompleted work, which in turn generated a very large number of LOIs, difficult to manage
- Completing the construction contractors' scopes of work with labour under the direct control of the owner could have given rise, in case of a defect or error causing an accident, to a dispute on the question of responsibility
- The operation went smoothly because an experienced individual came on the site armed with a proven methodology and the necessary software, and was given the authority to get on with the job
- The owner achieved his objective of a rapid turnover of the whole plant, which led on to successful commissioning and start-up, but it is still worth noting the real and potential difficulties cited above.

Appendix A: Civil Works and Buildings

The civil works construction contractor is the first to arrive on the site. Most of its work will have been completed before contractors of other disciplines can start, for civil works create the physical infrastructure upon which equipment is erected and piping and cabling supported. The principal types of civil works are:

- Surveying
- Earthworks
- Piling
- Underground drains
- Reinforced concrete for machine plinths, foundations and structures
- Roads and car parks

This appendix also briefly evokes building works. Structural steelwork is sometimes considered to be part of the civil works scope, but in this handbook it is treated as a distinct discipline (see Appendix E).

The parties involved on site can include the following:

- The EPCM contractor
- The architect
- The civil construction contractor
- Possibly a separate building construction contractor
- Building subcontractors covering the different trades
- Suppliers of concrete mix
- Special-service providers for surveying and for laboratory and on site testing facilities for soils and concrete

Early in the project a survey of the site will have been carried out to determine its shape, dimensions and levels in relation to survey monuments, the latter being tied back to a national coordinate system and/or to features of an existing plant in the case where the project is an extension. This survey will be recorded on a site plan, which will be the starting point for the determination of the layout of the plant and the subsequent detailed design. During construction it is important to protect the survey monuments and to regularly verify and record their positions. When the site is opened, the surveyor's initial task will be to establish alignments, coordinates and levels and battery limits for the various process, utility and off-site areas as well as those of access roads and car parks. From then on it will be necessary to set out positions of piles, drains, foundations, equipment plinths, structures and buildings. Accuracy of positioning of anchor bolts will facilitate equipment erection to close limits, which in turn will enable interconnecting piping to be realized with a minimum of mismatches. The axis of a pipe rack or of a pipeway can be a reference from which adjacent equipment can be positioned. Records comprise survey reports and marked up drawings.

The design and construction methods to be used for foundations for equipment, structures and buildings and of the associated roads and car parks, and also for earth retaining walls are dependent on the nature and physical properties of the soil on and under the site. For this reason the following factors will have been investigated by means of *in situ* and laboratory tests carried out by specialists to determine:

- The nature, thicknesses and stratification of the various layers
- Their properties, in particular their load bearing capacities, information necessary for deciding the type of foundations to be provided, in particular whether or not piling will be required
- The presence and extent of groundwater

On-site earthworks include:

- Removal of unsuitable or excessive soil
- Temporary storage of soil

- Earth moving, levelling and grading to create platforms for process units, off-sites, utilities and infrastructure, with compaction per lift at optimum moisture content, respecting permitted lift heights
- Excavating for drains, foundations and equipment plinths, including reinstatement and compaction
- Drainage, de-watering, shoring-up and temporary works

Once the platforms are completed and consolidated, battery limits, roads, foundations and plinth positions can be set out. Quality records comprise topographical survey reports, soil test reports carried out at various stages of the works and marked-up drawings.

Depending on bearing characteristics of the soil, strata depths and prospective loadings on foundations, piling may be necessary. There are many varieties of pile and piling techniques available. However, we can divide them into two broad categories:

- Bored piles
- Impact driven piles

Whatever the type of pile used, an initial survey must be made to determine pile locations. Care must be taken to position the pile accurately and a survey will be required afterwards to determine the coordinates as installed. Prior to the piling campaign there may be type tests carried out *in situ* on a small number of piles to determine resistance to loading. The results of these tests must be duly recorded.

A bored pile is by definition formed by boring, removing the earth and generally ensuring that the hole does not collapse by the insertion of a temporary steel liner. Special arrangements may have to be made in the case of the presence of ground water. A steel reinforcing cage fitted with spacers is positioned in the bore. Concrete is then introduced according to a predefined procedure, the steel liner being withdrawn progressively as the level of concrete rises. The inspection and test plan should include checking the boring rig position, for the reinforcing cage the grade of steel against mill certificates, the diameters and configuration of the re-bars, spacers for concrete cover, and the quality, placement and traceability of the concrete.

Driven piles can be in different formats and materials. Should the piles be in pre-stressed concrete, their prefabrication needs to be closely supervised. Besides concerns for the quality of the concrete, very close control must be exercised in tensioning the steel wires and in subsequent stress transfer to the concrete. Formal release for handling and subsequently driving the piles can only be given when crushing strengths of the corresponding concrete samples have reached or exceeded a predetermined value. Each pile should be identified with a serial number tied back to its quality records. Pile driving has to be carefully monitored and documented. Noting on the same form the designation of the pile geographical location and the pile manufacturing serial number enables traceability to be maintained between the operations on the prefabrication yard and those on the site. The pile is pre-marked at specified intervals along its length to enable the number of hammer blows between intervals to be counted and recorded as an indication of the resistance as the pile is driven deeper. Where a cluster of piles is driven, there may be interaction between them giving rise to heave. It may be necessary to re-drive one or more of the piles.

Underground drains may be simply for rain water, but there may be other separate self-contained systems in process areas to contain dangerous fluids in case of a leak or a spillage, possibly mixed with fire fighting water. A number of different materials can be used for pipes and fittings for underground drains, in particular, reinforced concrete, steel pipe welded and wrapped, GRP or PVC. Special bedding and backfill material may be specified so as not to damage the drains. Close surveillance will be required during all stages of the works, and so an inspection and test plan needs to cover:

- Prior verification of materials
- Excavation, bedding, backfilling and compaction
- Laying, alignment and jointing between pipe lengths and with the manholes
- Hydrostatic testing
- Surveying at all stages for level, grade and alignment

Reinforced concrete can be used on site for the realization of such works as:

- Equipment plinths at ground level
- Structures for supporting elevated equipment
- Earth retaining walls
- The whole or part of pipe rack/pipeway structures
- Access bridges across pipeways
- Underground ducting
- Paving for certain process areas
- Roads and car parks
- Primary structures of buildings

The constituents of concrete are:

- Fine aggregates
- Coarse aggregates
- Water
- Cement
- Additives (In response to ambient conditions or to produce a desired effect)

These are brought together in a batching plant, which may be already existing external to the site or may be erected on site for the duration of the works. Prior to starting the concrete works, the plant should be audited to ensure that it is able to meet project requirements. This involves reviewing the operation of the plant and its vehicles, as well as the sources and characteristics of the raw materials. Trial mixes must be made and tested. Including the provision of premixed concrete within the scope of work of the civil works construction contractor is preferred as it means that there is a single organization responsible for quality of concrete from raw materials, through batching, placement and curing to finished structure.

On the site itself pre-pour investigations include checking the steel grade and diameters of reinforcing bars and their configuration, the dimensions of anchor bolts and their positions, checking formwork, including reinforcing and supports and checking spacers to ensure adequate concrete cover of reinforcing bars. These verifications must lead to a formal written authorization for pouring the concrete.

As each delivery truck arrives at the site where the concrete is to be poured, a slump test is made to verify the plasticity of the mix. Water may be added within specified limits. Concrete placement shall be closely supervised and recorded. Samples in the form of cylinders or cubes shall be taken at a specified rate, duly cured and tested for crushing strength after 7 and 28 days. Since test results are only available long after the completion of the works, the operation must be fully traceable so that in the event of the values obtained being less than the specified minimum, the location of the corresponding concrete works can be located for investigation and possible corrective action. Traceability must also extend via the concrete delivery ticket back to the batching plant and to the raw materials. Where possible the special-service provider which is going to carry out the crushing tests can with advantage be made responsible for taking the samples at the pour site, marking the serial numbers and dates and for supervising curing both as to method and duration. Figure A1 illustrates how traceability of reinforced concrete can be achieved.

Curing needs close supervision both in terms of method and duration. Clearance to strike the form work must be recorded as must subsequent inspection, survey, repair, if any, and markings of the finished reinforced concrete structure.

Roads and car parks will be specified in terms of the type of material and thickness of each layer up to the wearing course, that is, the finished surface. The starting point is a well compacted sub-base accurately surveyed to ensure level, grade and alignment. Each layer must be compacted respecting the limit of lift height and verified by tests carried out at selected points. If the wearing surface is to be reinforced concrete slabs, then the principles proposed above in this appendix must be followed, with care taken over the joints made between adjacent slabs. Should asphalt be specified, the batching plant needs to be audited, materials and mixes verified and arrangements for delivery defined. Curbing and gutters need to be set out to take account of the wearing layer levels and grades. Quality records include certificates for the materials, tests results, survey reports and drawings marked up 'as built'.

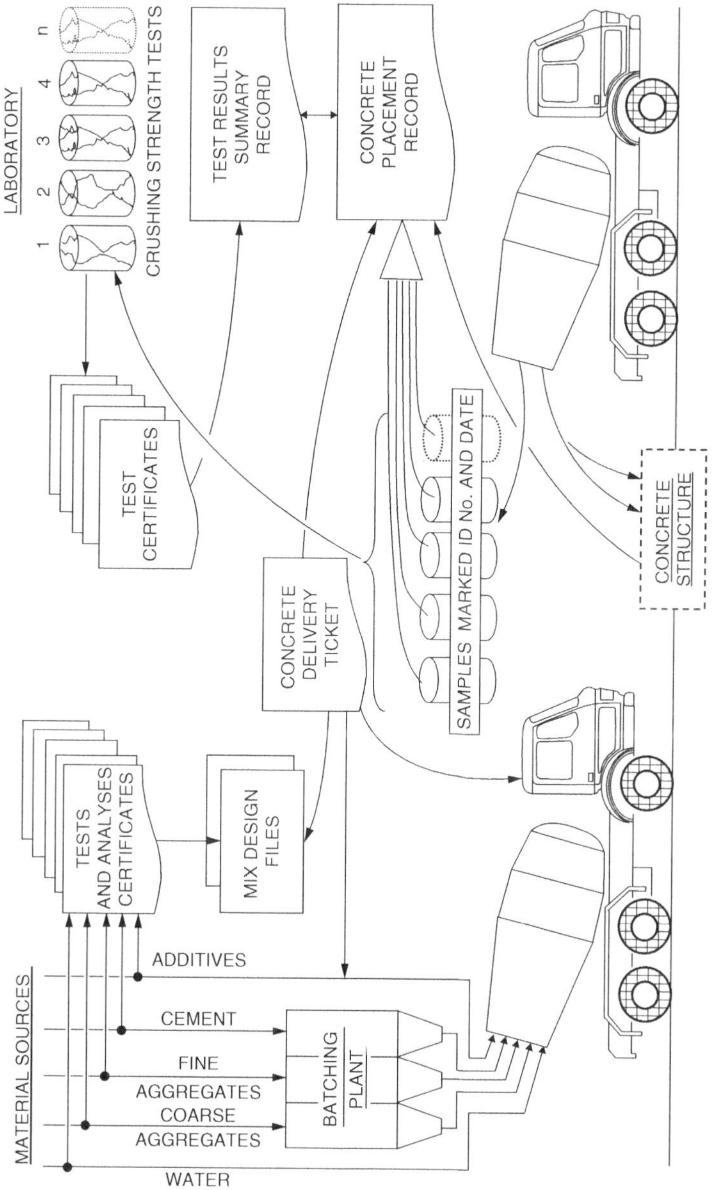

Figure A1 Concrete traceability.

Process buildings may include heavy reinforced concrete structures to support items of equipment and to carry the rails for overhead travelling cranes, such buildings in effect falling within the domain of 'civil works'. Other buildings such as offices, stores, workshops, the canteen, the infirmary, cloakrooms, toilets, control rooms, electric substations and gate houses, already evoked in Section 3.10, are more logically included in a contract awarded to a specialist building construction contractor. The latter may well carry out detailed design of the buildings and will be responsible for the coordination of the various sub-contractors representing the different trades. A number of proprietary items such as windows, doors, locks, sanitary ware, sheeting, insulation, panelling, suspended ceilings, floor and wall coverings, etc. will be incorporated into the works. It is useful to set aside a display area where samples of these items can be reviewed thus identifying potential problems early rather than when the works are well advanced. These reviews can be formalized as hold points. Buildings are generally subject to building codes and regulations; this requires ongoing vigilance and coordination with the Authorities concerned. An inspection and test plan is needed to be applied as the work of the various trades advances, in particular including hold points to be implemented at intermediate stages before the work is covered up.

Appendix B: Mechanical Equipment

This appendix outlines the major factors which have to be taken into account from the quality point of view for the erection of mechanical equipment, the more common items of which can be classified into one or other of the following broad categories:

- Unfired pressure vessels
- Boilers and furnaces
- Cooling towers
- Site erected storage tanks and spheres
- Rotating machines
- Lifting and material handling equipment

In addition to the appointed mechanical construction contractor, there will be present on site suppliers' representatives for certain categories of equipment in order to supervise the more critical phases of erection and testing.

Items of equipment can be set on a concrete plinth or on a steel structure. Taking the first of these cases, the concrete structure upon completion of curing and of a survey 'as built', will be released for setting the item of equipment for which it is intended. The top surface must be roughened and anchor bolts sealed. The item of equipment or the base plate of a driver/driven set must be positioned axially, transversally, vertically and to level. It is important to check the position and alignment of any flanges to be connected to piping. Instructions of the equipment supplier and of the grout manufacturer must be closely followed for mixing, placing and curing of the grout and tightening of the anchor bolts. Depending on the application, test cubes may be taken to be cured and then used for a crushing strength test of the grout. Some machines, instead of being mounted on a base plate, are positioned directly on metal sole plates previously grouted to the top of the plinth. Setting and levelling an item of equipment on a steel structure is simpler, being essentially shimming and bolting to the support pads so that the load is distributed. Depending on the application a base plate may be attached to the structure by welding.

A typical process plant can include a large number of pressure vessels completely manufactured and tested in the Supplier's premises and delivered on site as a complete finished item. Exceptionally a large vessel may be delivered in two pieces to facilitate transport and then be welded together on site. This operation would almost certainly be carried out by the supplier's personnel. For pressure vessels the requirements of the concerned authority will be very much to the forefront. There will be a quality file established in the country of manufacture with records of materials used, manufacturing processes, in particular concerning welding, with traceability, and the records of inspections and tests carried out. Cold-stamping on the vessel itself and on the name/data plate may be an indication of its conformity and acceptance status.

Erection of pressure vessels is relatively straightforward. Once preliminary checks have been made, the vessel will be positioned, aligned, shimmed and, where applicable, grouted, with special attention being paid to nozzles positions and orientation. Where there is a sliding foot contact, it must be positioned, aligned and lubricated. Anchor bolts should be tightened to specified requirements. Access platforms, hand railing and ladders (if any) can be added. These are generally supplied within the vessel purchase order. Where appropriate an internal visit will be made to inspect for cleanliness and to install/check any accessories, which can include items such as baffles, trays, and weirs. In certain cases the internals require setting up and levelling to close limits. Site pressure testing can be for the vessel as such, for nozzle reinforcings, for internal pressurized items such as steam heating coils and in the case of heat exchangers, tests for both the tube and shell sides.

For a packaged boiler delivered in one piece, depending on authority requirements, it will normally need to be pressure tested after installation. There will be accessories to be installed such as the chimney, exhaust ducting, burner, motor fan set(s), gauge glasses, instruments, platforms and ladders, relief valves and dampers, with refractory linings to be completed.

Concerning boilers and furnaces which are delivered as subassemblies and erected on site, the scope of work includes erection of supporting steelwork, erection and welding together of pressure retaining elements, erection of steel casings and the realisation of refractory linings using refractory bricks or refractory concrete. These are complex works normally carried out or at least closely supervised by a specialist design and build contractor, with inspection and test results entered on a whole family of record forms. Welding of the pressure retaining systems is to be the subject of rigorous and documented control, followed by pressure testing on completion. Refractory works must be based on a procedure defining preparation, mixing, placement, inspection, testing, repairs, curing and drying.

Cooling towers in a process plant project may be required in connection with:

- Air-conditioning
- A process requirement
- Power generation

The simplest are the smaller prefabricated units, which do not require much more than to be set in place, and connect to the piping and electrical circuits. The larger units built *in-situ* are generally the work of specialized design and build contractors. They may be naturally ventilated, or more usually of the forced or induced ventilated type. They can be of several cells grouped together, each cell comprising a basin at ground level, a concrete or sometimes wooden containing structure, a fan for upward air movement, a spray system with pipe work and nozzles as well as filler material over which the water will cascade and disperse to ensure maximum heat exchange between the water and the air. The basin will have a make up water connection to compensate for evaporation. Any concrete works can be dealt with in a classical manner as described in Appendix A. Obviously, water proofing is an essential element in the construction of the basin and the enclosing structures. An induction fan situated on the top of the tower is usually mounted with its shaft axis vertical, either belt driven or connected through a gear box and shaft to the drive motor. The rate of air flow and hence the load on the motor can be adjusted by changing the blade angle.

Only the smallest storage tanks can be delivered and erected in one piece. More usually a tank for atmospheric or low pressure duty will be welded *in situ* from previously cut, edge prepared and where appropriate rolled plates and prefabricated subassemblies. Some tanks are designed with a floating roof. The first success factor is the careful planning, execution, supervision, inspection and testing of plate preparation and prefabrications, with formal clearance to be required before shipment to site. The prefabrication shop should be the subject of an audit before any work begins (see Appendices D and E). A shop inspection and test plan supported by a complete set of record forms must address the following issues:

- Inspection of incoming material traceable to material and test certificates
- Definition and documentation of welding, that is to say, weld procedure qualifications, procedure qualification records and welder qualifications
- Nondestructive examination and destructive testing
- Execution of plate cutting, rolling and edge preparation, and of prefabrications (manholes for example)
- Markings for traceability and to facilitate erection
- Packing for shipment to site, in particular providing cradles to support rolled plates

The above gives rise to a set of inspection and test records, leading to formal release before shipment to site.

Concerning the site erection works, there are a series of successive stages to be defined in the inspection and test plan with a set of model forms adapted to the specific needs of the tank concerned. The following quality issues must be addressed:

- Formal clearance for the foundation, complete, surfaced and equipped with any under floor drains and connections, before the floor plates are placed
- Inspection of incoming material, traceable to documents
- As cited above concerning prefabrication, definition and documentation of welding requirements as well as those of non-destructive examination and destructive testing
- For the floor, then for each successive ring, terminating with the roof, execution of fit-up, tack welding, welding, inspection and testing of welds, repairs, dimensional and geometric checks and mark-ups to ensure traceability
- Erection of stairs, ladders, handrails and any other accessories
- Hydrostatic test and settlement measurement, clean and drain
- Review of complete quality file, fixing and stamping name/data plate
- Blast/paint/insulate to specified requirements

The same general considerations apply for the site erection of spheres, with an added emphasis on accurate preparation of plates and careful fit up of plate edges prior to welding.

In process plant the most common driven machines are pumps, compressors and fans coupled to electric motors, but prime movers – gas and steam turbines – as well as diesel engines can be used for certain applications. The simplest example is that of a small centrifugal pump directly driven by an electric motor, mounted on a common base plate. At the other extreme we find for example a large compressor driven by a gas turbine through a gearbox, with air intake trunking and filter, exhaust trunking, silencer and stack, and complex auxiliary systems for fuel, lubrication, cooling and shaft sealing. Whatever the size of the machine, certain basic erection activities and corresponding inspection and test points, are broadly speaking realized respecting the same general principles. The machine has to be set, and where applicable grouted, following the principles evoked above. Driver and driven can be positioned on a common base plate, but for some large machines, each is positioned on previously grouted sole plates.

The simplest form of coupling is rigid with driver and driven shaft flanges bolted up in contact face to face, an arrangement typically used for smaller machines. For larger and more sophisticated drives, in particular where high speeds are involved, the shaft flanges are spaced apart to allow the fitting of a flexible coupling. In aligning the shafts and fitting the coupling, the manufacturers' instructions must be scrupulously followed. During transport and the initial stages of erection the coupling will be broken, but once the machine is installed on its foundation, a preliminary shaft alignment will be carried out. Then a second alignment is made with the associated piping in position, with the piping flanges and machine flanges parallel, touching with gaskets interposed, with bolting sets in place, but not yet tightened. With dial indicators in place to check shaft alignment, the flanges bolts are tightened and at the same time any change in shaft alignment is noted. The piping is adjusted until the change of shaft alignment between the loose and bolted up state is within specified limits. At the same time as the alignment, the axial distance between shaft flange faces has to be adjusted within specified limits. For some machines, a deliberate axial mismatch is introduced to compensate for thermal expansion. A further set of readings will be taken during commissioning to verify alignment when the machine is hot. At the same time as carrying out shaft alignment, a 'soft foot' check should be made on the driver and driven machines, to ensure that load bearing is shared by all contact areas between the machine frames and the base or sole plate. If specified, driver and driven machines may be dowelled in place once final alignment is attained.

Other actions concern the lubrication of bearings and installation of shaft seals. However, for larger machines, more complex systems for lubrication, cooling and shaft sealing are provided with auxiliary equipment often mounted together on a separate skid. These auxiliary systems require careful preparation and checking to the requirements of the supplier. This is generally carried out under the supervision of the supplier's representative.

Where driver and driven are connected by belting, the first verification in the case of a multiple belt drive, is to ensure that the belts are a matching set. Electric motor direction of rotation is checked uncoupled. Then as the belts are tensioned, alignment is checked by means of a straightedge and feeler gauges on the pulley faces. Concerning belt tensioning, the supplier's recommendations must be followed.

Lifting equipment can include:

- Overhead, gantry and under slung travelling cranes
- Monorail hoists
- Pedestal cranes
- Passenger and goods lifts

In a process plant some of these may be directly linked to production, whilst others may fulfil a support role, for example a monorail hoist positioned over a machine to facilitate erection and maintenance. They will almost certainly be the subject of close scrutiny by the competent authority of the country in which the plant is situated and in the country of manufacture. So it is in everyone's interest to make sure that regulatory requirements are know and followed from the initial design stage onwards. Lifting equipment will have been wholly or partly erected and tested in the factory and will generally arrive on site as sub-assemblies, rarely as a complete unit. Stressed items such as hooks and wire ropes will have been individually proof load tested by their suppliers. This means that there will be a quality file which will arrive on site with the equipment, to be further documented to record inspection and tests carried out during and upon completion of erection. Whilst smaller simpler items of lifting equipment may well be erected solely by mechanical construction contractors, for more complex ones, in particular for lifts, erection and testing will be closely directed and monitored by the supplier's representative.

We do not have the space to discuss all the different possible lifting devices available. Instead let us take as an example the case on an overhead travelling crane. Quality concerns will focus on:

- The down shop rails, their fixing, alignment and levelling as well as the end stops
- The alignment and bolting up of the main girders square with the end carriages and the positioning of the crab on the cross shop rails

- The hook-up of the electrical installation on the crane itself and connection to the down shop power feed system with the basic electrical checks
- No load tests, hook approaches and stops/limit switch adjustments for all motions in both directions
- Safe working load test including measurement of structural deflection
- Overload test followed by visual examination of stressed zones and components
- Final review of documents and acceptance by the authority concerned

Material handling systems for dry bulk solids may be based on moving discrete lots in containers, or by an arrangement giving a continuous throughput. This can involve a wide variety of equipment many of them proprietary items. Bulk material is stored in silos or hoppers loaded from above, with the bottom tapering to an outlet fitted with a feeder to regulate outwards flow. For vertical lifts use can be made of bucket conveyors. Screw conveyors can be used for short distances on the level or inclined. Transfer of material in a horizontal plane or within limits sloping upwards is the role of the troughed belt conveyor, a typical inspection and test programme of which could include:

- Surveying foundations
- Checking alignment of support structures
- Inspecting installation and alignment of driver unit, tensioning unit, belt scraper unit, and idler pulley
- Inspecting and testing electric motor, controls, cabling and earthing, checking motor direction of rotation
- Inspecting installation and alignment of trough idler rollers and return idler rollers
- Checking belt tension
- Testing safety trip

Appendix C: Heating, Ventilation and Air-Conditioning

In a process plant project the extent and configuration of heating, ventilation and air-conditioning (HVAC) will depend on the processes, the buildings, manning and on the site climatic conditions. A project situated in the tropics can be very different from one realized in a temperate climate. This is a detailed and dense technology, with many options and many variations on each option. In this appendix we shall simply evoke the more common kinds of applications and identify quality issues which have to be addressed.

The following are typical contractual arrangements:

- The EPCM contractor designs the system down to the last detail, procures all of the materiel and then engages a specialist construction contractor for the site work.
- The EPCM contractor carries out basic design, procures only major items of equipment and then engages a construction contractor to do the detailed design, to procure outstanding materiel and to carry out site erection.
- The EPCM contractor defines only basic parameters and then requires the chosen construction contractor to do a complete design, procure all of the materiel and carry out the site work
- The last option above is incorporated into a global contract to design and build a complete building, so that the HVAC specialist firm is a subcontractor of the construction contractor.

Depending on specific requirements, work on site can include some or all of the following:

- Installation of individual items of equipment such as fans, cooling towers, compressors, boilers
- Installation of packages such as an air handling unit or a refrigeration unit
- Erection and connection of ducting with necessary supporting steelwork and installation of in-line items such as dampers and filters
- Installation of instrumentation and control devices together with their interconnecting wiring
- Inspection and testing to ensure correct installation and functioning

This prepares the ground for commissioning of the whole installation.

We shall consider in turn heating, ventilation and air conditioning, briefly evoking the types of application to be met and the corresponding materiel to be installed. Then we shall examine quality concerns to be addressed during erection.

Depending on requirements, a local or central heating system may be provided. Typically, in an industrial context, local heating can be carried out by installing in the space to be heated one or more self contained fuel oil or natural gas fired units comprising:

- A burner and combustion chamber
- A heat exchanger through which pass the products of combustion to discharge outside of the building
- A fan recycling air within the building with a proportion of air drawn from outside, delivered through the heat exchanger secondary circuit, then to be projected through diffusers into the space to be heated

The advantage of such units is that they are completely assembled and tested in the manufacturer's works, so site installation and its associated inspection and testing are minimal.

Central heating means generating heat by combustion of fuel in one place and distributing it to the places where it is needed by means of a canalized fluid. This can be air in ducting connected to an air heater unit generally similar to the one described above, or alternatively through the use of hot water or steam generated in a boiler and conducted through piping to areas to be heated. It may be that steam is available on site anyway, and so it can be used for heating without the need of a specific heating boiler. The heat in the piped steam or hot water can be transferred to the air in the space to be heated by radiant panels or by convection. Heat transfer can be amplified by the presence of fans in the heating units.

Natural intake and exhaust can occur through louvred openings fitted in the building walls and/or on the roof. These are proprietary items designed to allow the passage of air, but preventing the ingress of rain and snow, the louvers being fixed or motorized. They can be supplied and erected under the building contract. The simplest way of mechanically ventilating a space in a building is by the installation of supply and/or extraction fans mounted in wall or roof units. The addition of ducting systems for supply and extraction ensures a more even distribution. When ventilation is needed in a specific location, where, for example, dust or fumes are being produced, a canopy can be suspended over the equipment concerned. The contaminated air is removed by an extraction fan and delivered outside the building through a ducting system possibly equipped with a filter or dust collecting unit.

It may be sufficient to heat and/or ventilate a building. However air conditioning takes matters a step further by maintaining the air in the concerned space at the chosen air temperature, humidity level and degree of air cleanliness, whilst evacuating heat generated by equipment and occupants and at the same time providing the necessary rate of air change. There are many different possible configurations. The simplest arrangement is that of a self-contained wall or window mounted unit suitable for air conditioning an individual room. The other extreme is a centralized system where the air after having been conditioned in an air-handling unit (AHU), is distributed throughout the air conditioned space by a system of ducting and then extracted through a second system of ducting so it can be recycled/exhausted out of the building.

The AHU contains supply and extraction fans to maintain the airflow, the latter being regulated by motorized dampers and louvres. Means are provided to heat or cool the air as well as to humidify or dehumidify it. The installation includes the necessary instrumentation and control elements with a panel housing the air conditioning management system.

Individual items of equipment such as fans, compressors, boilers and cooling towers can be installed following the guidelines set out in Appendix B. The inspection and test programme for an AHU could cover:

- Name/data plate details
- Positioning and levelling, anchor bolts and anti-vibration mountings
- Connections to ducting, gaskets in place, flanges bolted up, seals in place
- Fans, transport stops removed, shaft free, alignment/couplings/belt tensioning/bearing greasing
- Filters elements in place, airflow direction
- Dampers and louvres, transport stops removed, movement free, airflow direction, operation
- Heating and cooling systems
- Humidifier and dehumidifier
- Electrical connections, earthing, insulation resistance, protective devices, fan motors direction of rotation
- Measuring, detecting and actuating devices in place, connected and tested

Ducting can be of rectangular cross-section made from galvanized sheet steel formed and riveted, strengthened by diamond break stamping, angled stiffeners and tie rods. Alternatively ducts may be of circular or oval cross section, and of other material than galvanized steel. Heavy duty ducting can be built up by welding steel plates to be surface treated on completion. It is important that the steel supporting structures are of sufficient cross section, properly spaced, and positioned to take account of the extra weight of in-line items such as dampers. Flexible ducting is included, in particular close to fans as a protection against transmission of noise and vibration. As a further precaution in-line noise attenuators can be included. Where the ducting connections are flanged and bolted together with gaskets, care must be taken to ensure that the flange faces are parallel, the holes are aligned and the gasket is correctly positioned to ensure an even contact as the bolts are tightened. For joints based on the use of mastic and heat shrink sleeves or special tapes, the supplier's instructions must be strictly applied. With the ducting in place, a thorough check must be carried out on the openings for access, for cleaning, for maintenance or for the introduction of instruments, to ensure that they are workable in practice, and to check that when closed they are properly sealed. Depending on the service in question, ducting may be required to be insulated. Formal inspection of the ducting system must be documented. A pressure or vacuum leak test may be specified.

Air control devices installed in ducting systems include manual dampers, motorized dampers and fire dampers. Their inspection and test plan should include:

- Name/data plate details
- Correct location, support, airflow direction
- Access for operation and maintenance
- Flange connections gasketed and bolted up
- Linkages, free movement, seating, clearances
- For motorized dampers, connections, controls, operation
- For fire dampers, certification, access to fusible link, simulation of operation

Static in line items, such as air supply outlets and air extraction inlets should be checked as installed covering name/data plate details, correct location, support, airflow direction and connection to ducting.

With inspection complete, loop tests can be carried out, leading on to starting up the HVAC system, measuring and balancing airflows.

Appendix D: Welding

On a process plant construction site, welding is an integral and essential part of:

- Prefabrication and site erection of pipework
- Prefabrication and possibly site erection of structural steelwork
- Prefabrication and site erection of storage tanks and spheres

In all these disciplines, weld integrity is closely linked to safety; so, in consequence welding has become codified into a highly defined system expressed in standards, which enables us:

- To define relevant welding procedures
- To prove or 'qualify' them
- To test and certify welders for their capacity to successfully apply the qualified procedures
- To define the conditions under which welding takes place
- To inspect and test welds according to defined acceptance criteria
- If necessary to repair defective welds and re-examine them
- To ensure that an acceptable level of traceability is attained
- To establish documents to demonstrate conformity to specified requirements

The parties involved are:

- The EPCM contractor's site supervision staff
- The construction contractors concerned, which provide welding equipment, welding consumables, welders and supervision

- The special-service provider (SSP), which carries out nondestructive examination and interprets the results
- The SSP which provide a laboratory service for destructive testing

There is a multiplicity of welding processes available with numerous variations. However, the following meet most of the requirements on a construction site, or in a prefabrication shop:

- Manual metal arc (MMA) or 'stick' welding using a coated electrode manipulated manually
- Metal inert gas (MIG) welding, with a hand held pistol bringing together an automatically fed consumable wire electrode with inert gas shielding
- Tungsten inert gas (TIG) welding using a nonconsumable tungsten electrode shielded by inert gas, together with a separate filler rod (when required)
- Submerged arc welding (SAW), for long continuous welds under a blanket of granulated flux

Welding operations on site or in a prefabrication shop are brought together to form a coherent whole by actions taken at three levels, each duly documented as follows:

- For each type of weld, full details of the weld and method to be followed are defined in a **weld procedure specification (WPS)**.
- For each WPS a sample weld is performed to be subjected to a range of nondestructive examinations and destructive tests, the results of which are entered on a **procedure qualification record (PQR)**, the whole operation being normally witnessed and certified by an independent third party. (If the tests meet specified requirements the WPS is 'qualified'.)
- A welder who will be called upon to weld according to a given WPS performs a corresponding sample weld, which is tested in a manner generally similar to testing carried out for the PQR. (If the results are accepted the welder is 'qualified' or 'coded', once again the operation is normally witnessed and certified by an independent third party.)

In practice the standards allow some flexibility in application within defined ranges of **essential variables**. At the design stage the aim must be to cover project welding requirements with a minimum number of WPSs. Then by a judicious choice of essential variables, a minimum number of PQRs can cover the whole range of WPSs. Similarly, welder tests can be defined such that a welder is qualified for a range of WPSs. This is a complicated subject whose management cannot be improvised, so it has to be the responsibility of a person familiar with the exigencies of the standards being applied. Qualifying weld procedures and welders is an expensive and time consuming exercise which has to be completed before welding can start. This is a key technical, planning and commercial issue to be clarified as early as possible. Under certain conditions, previously established PQRs and welder qualifications may be acceptable, which can show a considerable saving in time and money.

Depending on steel grade, weldment thicknesses and configuration, pre-heating and interpass heating may be specified. This complicates the task of the welder, giving rise to the temptation to cut corners, so close supervision is advisable. For some items post-weld heat treatment for stress relieving may be required to be applied. This calls for uniform heating to a predefined temperature/time profile, with a temperature rise phase, maintenance at the plateau temperature for the required duration and then cooling at a specified rate.

The expression 'nondestructive examination' (NDE) is a generic term encompassing visual inspection and the following techniques:

For surface inspection

- Dye penetrant inspection (DP)
- Magnetic particle inspection (MP)

For detecting internal discontinuities, volumetric techniques

- Radiography (RG)
- Ultrasonics (US)

DP is often carried out by construction contractor personnel, whilst the three other techniques are generally executed by staff of SSP. The latter may be engaged by the construction contractor as part of the contract scope or alternatively directly by the EPCM contractor as an independent third party, in particular to meet authorities' requirements. The EPCM contractor must ensure for every area of the welding programme that:

- The type and level (e.g. 10% or 100%) of NDE are predefined
- Methods, procedures and acceptance criteria are predefined according to appropriate standards
- Personnel provided to carry out NDE and interpret the results possess qualifications tied back to a standard

Visual inspection is the primary verification carried out on the surface of all welds. Visual inspection can be enhanced by DP and MP to reveal surface discontinuities not visible to the naked eye. DP requires the successive use of several fluids. It is important to follow precisely the product supplier's instructions if DP is to reach its full potential. MP is by definition only utilizable on ferromagnetic materials. It depends on magnetizing the item under test then applying a ferrous powder or a fluid with ferrous powder in suspension. Discontinuities at the surface, or just below, are revealed by patterns developed by the magnetic field. For both DP and MP, defects should be marked, and photographed or noted on a drawing.

Coming now to volumetric techniques the choice is between RG and UT, though it is not always a free choice. Certain types of weld and certain types of defect lend themselves more to one than to the other. In addition, traditionally RG has been favoured with UT in a relatively marginal position, but this situation is evolving. Should the choice between using UT or RG be left open subject to mutual agreement, it is important that the question be resolved as soon as possible. With RG, a film gives a permanent record; whilst a digital system permits viewing on location and image manipulation. However the safety aspects necessitating shutting off the surrounding area and evacuating personnel is a complication which does not exist with UT. RG can be carried out with either X-rays using a suitable apparatus or by gamma-rays from a radio active source. Both techniques possess advantages and inconveniences in terms of quality of image and convenience in application on a construction site. However, use of a radioactive source suffers from the major disadvantage of requiring permanent protective measures, whether it is in use or not, and furthermore, can give rise to administrative difficulties, especially when crossing frontiers.

Defects revealed by visual inspection, possibly enhanced by DP and MP, include cracks, undercutting, surface porosity, surface slag inclusions, craters, residual slag, incorrect weld reinforcement profile and misalignments. Defects detected through the weldment thickness by volumetric techniques can include cracks, lack of penetration, lack of fusion, porosity, slag inclusions and laminations.

Standards go to great lengths to define acceptance criteria, in particular the limiting conditions beyond which the defect is not acceptable and hence has to be ground out and repaired. Since the standard cannot cite every conceivable defect configuration and since certain defects are by their nature difficult to assess, there is inevitably at times an element of subjectivity, which enters into the decision to repair or not. The answer to this dilemma is to use for these tasks highly skilled and experienced individuals independent of production. It is desirable to analyse regularly, say weekly, the numbers, percentages and types of welding defects and the performance in this respect of individual welders. These can be presented in the form of graphs, histograms or pie charts. Any trends for the worse can be identified early and corrected before they become critical. Computerization of the quality records can facilitate collation and presentation of this information.

Destructive testing of a weld sample is carried out as part of a weld procedure qualification, of a welder qualification test or of a production test. The welded test plate will be machined and ground to produce a number of samples either longitudinal or transversal to the weld depending on the test to be performed. The range of verifications includes tensile tests, bend tests and impact tests on notched specimens as well as macroscopic and hardness survey of a cross section including the weld itself, the heat affected zone and the parent metal. These tests are carried out by qualified staff in a laboratory equipped with machining, grinding and polishing facilities and the necessary test equipment. It is important to be able to obtain test results quickly. This is generally not too difficult in the industrial areas of developed countries, but in some parts of the world this can be a problem with samples having to be sent long distances to a laboratory where they may receive little priority. This matter is already mentioned in Section 3.6, which evokes the advantage of establishing laboratory facilities on the site. Test record forms are usually those of the laboratory.

Traceability has already been discussed in general terms in Section 7.6. It is particularly important in the welding discipline, but the level to which it is pursued is related to the works in question. For example, for a low-pressure potable-water system, traceability would be minimal, whilst for say piping, part of a high-pressure steam line, it would be complete. Full traceability means establishing a trail between the weldment in question, heat numbers of the component parts and of weld consumables for each shop and site weld (leading to the corresponding certificates), the WPSs/PQRs, welder's identification and qualification records, welding records, NDE reports and reports of repairs. The management of traceability can be computerized using a spread sheet or by marking up the details on suitable drawings or diagrams. These could be specially formatted to facilitate the recording of traceability information or marking up could be made on existing drawings, such as isometrics for piping or general arrangement drawings for structures. These drawings or diagrams are called 'weld maps'. Depending on code requirements, the weldment itself may be required to be cold stamped with certain of these elements of information, such as the number identifying the welder concerned. Loss of traceability on a critical item can be difficult and costly to restore. This would give rise to a special audit (Chapter 9).

A smooth welding operation giving consistently a level of quality to meet specified requirements does not happen on its own. It is instead the result of scrupulous attention to detail both in preparation and execution. The points requiring attention are:

Before welding starts

- Presence of competent supervisors
- Availability as required on-call of SSPs for NDE and destructive testing

- Availability at the work face of specific detailed work orders, with relevant drawings, specifications, standards, WPSs, a complete inspection and test plan, as well as quality record forms ready for use
- Traceability arrangements, clearly defined and understood by those who have to ensure that they are in effect carried out, with formatted weld maps available
- Availability of equipment and consumables for all stages of welding and verification, that is, welding sets, hot quivers for electrodes, inert gas supply, heat treatment facilities and inspection and testing apparatus
- Identity of welders, verification of their certificates and badges
- Verification that material both sides of the welds and weld consumables correspond to specified requirements and are traceable to test and analysis certificates
- Work environment, that is, access, position of welding, shelter, lighting
- Edge preparation, fit-up and tack welding

As welding progresses

- Respect of the weld parameters such as pre-heating, current intensity, speed of advance, number of passes
- Intermediate inspections, if required, duly documented
- Removal of slag between passes
- Repair and re-inspections if required

On weld completion

- Visual and other NDE
- Repair and re-inspection if required
- Post weld heat treatment, if required
- Marking, traceability, quality records
- Final inspection

Appendix E: Structural Steelwork

Early in the design phase of a project a decision will have been taken regarding primary structures. Will they be in reinforced concrete or in steel? Such structures can include:

- Pipe racks and pipeways
- Supports for elevated fixed equipment
- Building main frameworks
- Support structures for overhead travelling cranes

If they are to be in steel, a common arrangement is for the EPCM contractor to carry out the basic design, and then engage a specialized construction contractor to do the detailed design, procure the material, produce the shop drawings, prefabricate and then to erect the structural steelwork on site.

Secondary structures can include:

- Access stairways and ladders
- Walkways and platforms
- Handrailing
- Continuous or grated flooring
- Supports for the outer skin of buildings

These items are generally defined as standard configurations in a specification with supporting drawings prepared by the EPCM contractor. They may be designed in detail, prefabricated and site erected as part of the contract for primary structural steelwork. However, secondary structures associated with an item of equipment, such as a column or a furnace, will normally be supplied and erected as part of that equipment. In the case where a construction contractor is responsible for a complete building covering all disciplines, steelwork prefabrication and erection may be the subject of a subcontract.

Site welding of structures with its attendant difficulties are where possible to be avoided or at least minimized. Instead, site erection should preferably, be based on the bolting together of prefabricated items. This implies prefabrication to close limits of dimensional accuracy with special attention being focussed on interfaces between:

- Column bases and anchor bolts
- Different structural prefabrications to be bolted one to the other
- Primary and secondary structural steel elements
- Structures and the equipment they support

Collation of information on these elements takes place at the design stage, to be incorporated eventually into the construction contractor's shop drawings from which the parts are to be made. Physical verification is an additional safeguard to avoid only discovering mismatches late in the day during site erection. This takes the form of presentation of templates and where appropriate, a trial erection at ground level. The latter is particularly important when a number of identical ensembles are concerned, such as repetitive building frames. The first set can be considered as a prototype to be assembled to check compatibility of interfaces and respect of over all dimensional tolerances. This can be defined as a hold point. Production of the remainder will only proceed once any problems have been identified and corrected.

Should prefabrication take place on or near the site, supervision may be within the work scope of the EPCM contractor's site team. The site quality plan and inspection and test plan established by the construction contractor would then cover both shop and site activities. Before work starts, the EPCM contractor shall ensure, preferably by formal audit, that the premises, equipment, skilled staff, supervision and procedures are such as to enable the contract to be fulfilled successfully. Special attention will be required if the prefabrication shop has been created from scratch, as opposed to one already operating as a going concern. Depending on circumstances the EPCM contractor may supervise prefabrication on a visit basis or by a full time presence. Appendix D gives a detailed discussion on welding, and in particular evokes the question of level of traceability and outlines a programme of surveillance. In addition to ensuring that requirements for welding quality are met, special attention must be focussed on:

- Accurate dimensioning and treatment of interfaces of structural members to facilitate erection on site (Are mating surfaces to be painted or not?)
- Trial erection, particularly in the case of the first in a series of interfacing subassemblies
- Blasting and painting if this is done at the prefabrication shop (see Appendix I)
- Identification markings corresponding to those on the erection drawings
- Lifting points marked as per the erection drawings
- Packing, slinging, dunnage and arrangements for transport to site

It is vital that no item should be allowed to leave the shop to be sent to site if it has not been submitted to and has successfully passed all the appropriate points in the inspection and test plan, leading to a formal authorization to ship. A rigorous programme of inspection and testing in the prefabrication shop pays dividends in terms of problems avoided on site.

Site erection must be defined in a timely manner with a written procedure prepared by the construction contractor and duly approved by the EPCM contractor. This procedure must take account of safety issues. Ideally, prefabricated items should be delivered in the order determined by the erection sequence. However, if they are to be stored on site, measures must be taken to ensure that they are adequately supported clear of the ground. Lifting points and slinging arrangements defined on the drawings must be followed to avoid damage during handling. The necessary upstream coordination with the civil works should have ensured that the anchor bolts conform to specified requirements and are accurately positioned.

Vertical and horizontal structural members are erected in the first place with temporary fastenings and supported by guy ropes to maintain stability until cross bracings are in place. When required, subassembly is carried out at ground level before lifting a complete ensemble into place. With sufficient members erected, measurements of verticality and horizontality can be taken and the structure adjusted accordingly. Values can be annotated on the drawings or on record sheets with the structure represented schematically. These values can be compared with tolerances permitted taking into account any accumulation of tolerances. Since structural integrity depends on the bolting together of the various items, special care must be taken to ensure that bolting sets are complete, are correctly dimensioned and are of the specified grade of high-strength (HS) steel traceable to material certificates. Torquing spanners shall be pre-calibrated to the specified value and thereafter checked at regular intervals to an established procedure. As each section of the structure is definitively bolted-up, for a specified percentage of bolts marked by the inspector, the value of torque which has been applied must be checked. Bolts in structures subject to vibration in service may be fitted with locking devices. HS bolts once tensioned cannot be reused. Should any bolts be found to be insufficiently torqued, more bolts must be checked according to a formula laid down in the specification. This exercise should be carried out by stages as the erection advances, so that any underlying torquing problems can be identified early and duly corrected, instead of it only being discovered with the structure complete. Column bases can be grouted as specified within the erection sequence.

Site erection should be recorded on suitable inspection and test forms and/or on marked up drawings upon which are entered details of measurements of levels, verticality and horizontality of structural members as well as details of bolted connections and of site welding, if any.

Appendix F: Piping

A process plant can include typically several tens of kilometres of piping. Although the discipline of piping, with its associated welding, runs within a tight framework set by standards and regulations, it is the discipline where there are potentially the most nonconformities and engineering queries. The reason for this is that it is in effect a gigantic puzzle built up from a large number of inputs. It is an extremely complex exercise to manage the design, procurement, shipment, storage, prefabrication, site erection, inspection, flushing, testing, air blowing, restoration, completion and turnover of the plant's piping systems. This environment calls for a clear definition of requirements and meticulous attention to detail throughout.

The parties concerned are:

- The EPCM contractor's supervisors
- The piping construction contractor(s)
- Specialist construction contractors executing multi-discipline contracts for complete units including piping, a water treatment plant for example
- Special-service providers for welding inspection and testing on site and in a laboratory

Figure F1 gives in the form of a flow chart an overview of the sequence of operations necessary to complete the construction of the piping in a process plant. For each stage there will be a programme of surveillance, inspection and tests with the completion of the corresponding records. Piping can only be released to the next stage if all the specified requirements have been met. Welding is an integral part of the prefabrication and site erection of piping, so this appendix has to be read in conjunction with Appendix D.

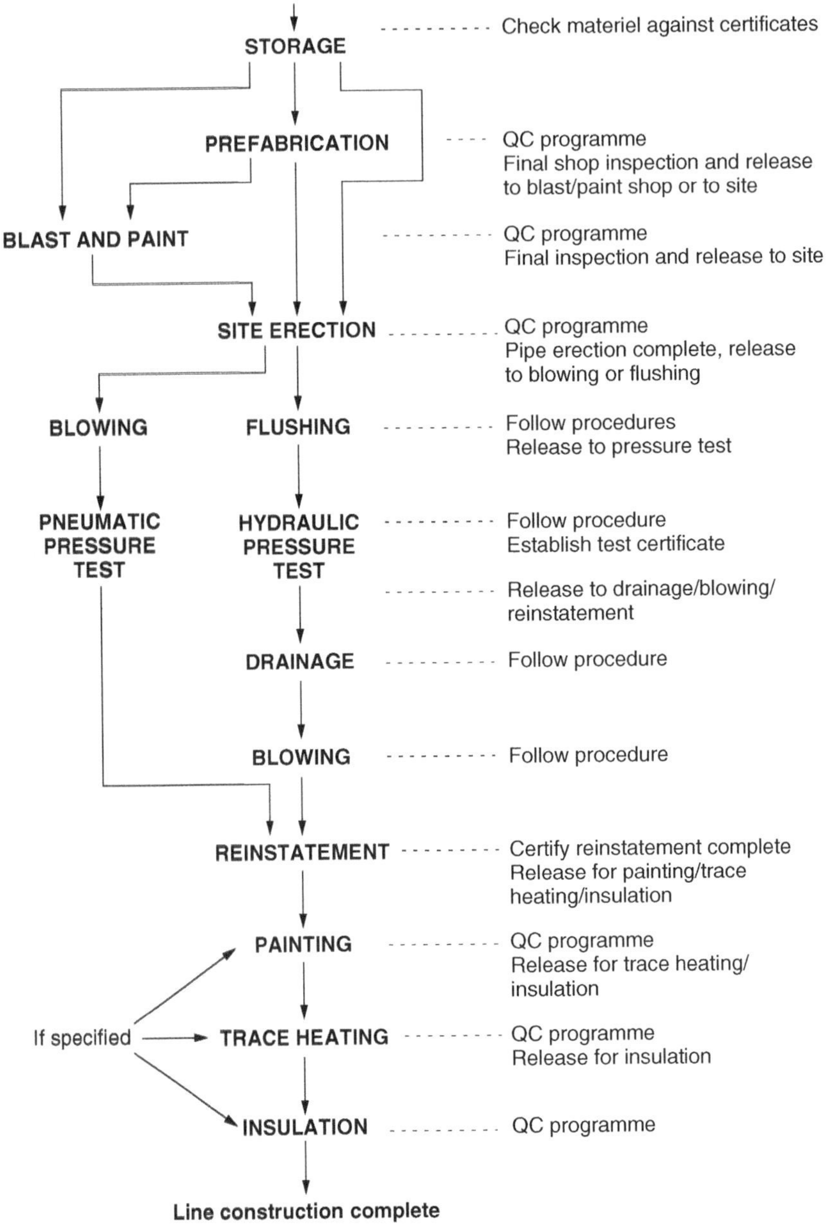

Figure F1 Piping works flowchart.

From its inception the design evolves as a reiterative process towards a definition of the piping systems presented as:

- Piping and instrumentation diagrams (P&IDs)
- Specifications to define piping materials for each range of duty
- Drawings and data sheets to define standard configurations
- Plot plans defining layouts of each process or other functional zones dimensioned to fix equipment centre lines and connection interfaces
- Isometric piping drawings (ISOs)

all of which are consolidated by alphanumeric identifications tied back to process units, functional systems, lines, and so on.

This may seem confusing, but we can conveniently visualize the ensemble of the piping of a process plant in terms of a hierarchy. Piping is first divided into **systems**, such as high-pressure steam, cooling water, instrument air, etc. According to its specific layout and spread, each system comprises a number of **lines**. All lines of all the systems are identified on a project **line list**, which states also the main characteristics of each line. The physical configuration of a line is detailed on a number of **ISOs.** Generally the finished physical piping corresponding to an ISO is too cumbersome to be transported and erected in one piece, so it is split into a number of **spools**, each comprising straight lengths of pipe and fittings as may be required joined together by **shop welds**. The finished spools covered by an ISO are shipped to site to be erected and joined to each other, to the piping of adjoining ISOs and to items of equipment, valves, and so on, by **site welds** or by flanged or screwed connections as the case may be. Each ISO includes a material list with quantities and generally other information including requirements for welding and heat treatment, together with an indication of applicable paint and insulation systems. The sum of the quantities of materials indicated on the ISOs can be integrated into a site wide list of materials to serve as a basis for procurement, shipping and expediting.

Chapter 7 already evokes the importance of managing in an orderly fashion the receipt, storage and issue of materiel of all types. These precepts are particularly pertinent in the case of pipes and fittings. Effectively managing the huge number of items which pass through the store is the first step to be taken for ensuring that prefabrication and erection will be achieved with a minimum of problems. Figure F1 already includes an indication of possible paths to be taken by piping material from leaving the store to being finally erected on site. This is shown in more detail in Figure F2. From this diagram it can be seen that the route to be taken for a given element of pipe depends on whether or not the line to which it is intended is to be painted, and if it is a straight length, whether it will go directly to the site or be cut up in the prefabrication shop to form spools. Whatever the route taken the pipe element must retain its identity for the purposes of traceability.

In Section 7.6 the principles of traceability are outlined and Appendix D shows in greater detail how these can be applied to welding. As far as piping is concerned the level of traceability must be defined contractually for each line class and clarified completely in detail at the outset of the piping contract.

Conditions on site can often be far from ideal, due to exposure to the elements (rain, wind, extremes of temperature), difficulty of access, the need to weld in unfavourable positions, etc. For this reason welding is carried out as much as possible in prefabrication shops to produce subassemblies thus minimizing site works. Should the prefabrication shop be on or near the construction site, the site team may be responsible for its supervision. Appendix E discusses prefabrication of structural steel elements. The quality concerns evoked therein apply in general to prefabrication of piping. Another area of prefabrication is that of pipe supports, which may constitute a wide variety of types and configurations.

Site erection of piping can be greatly facilitated if the items of mechanical equipment interfacing with piping, through flanged, screwed or welded connections, are accurately positioned in conformity with the plot plans, the survey being duly recorded. A decision can then be made to what extent to fabricate spools exactly as per the ISOs, or alternatively to include excess lengths on chosen spools to be offered up and exactly cut to length on site. Flanged piping interfacing with items of equipment, in particular with rotating machines,

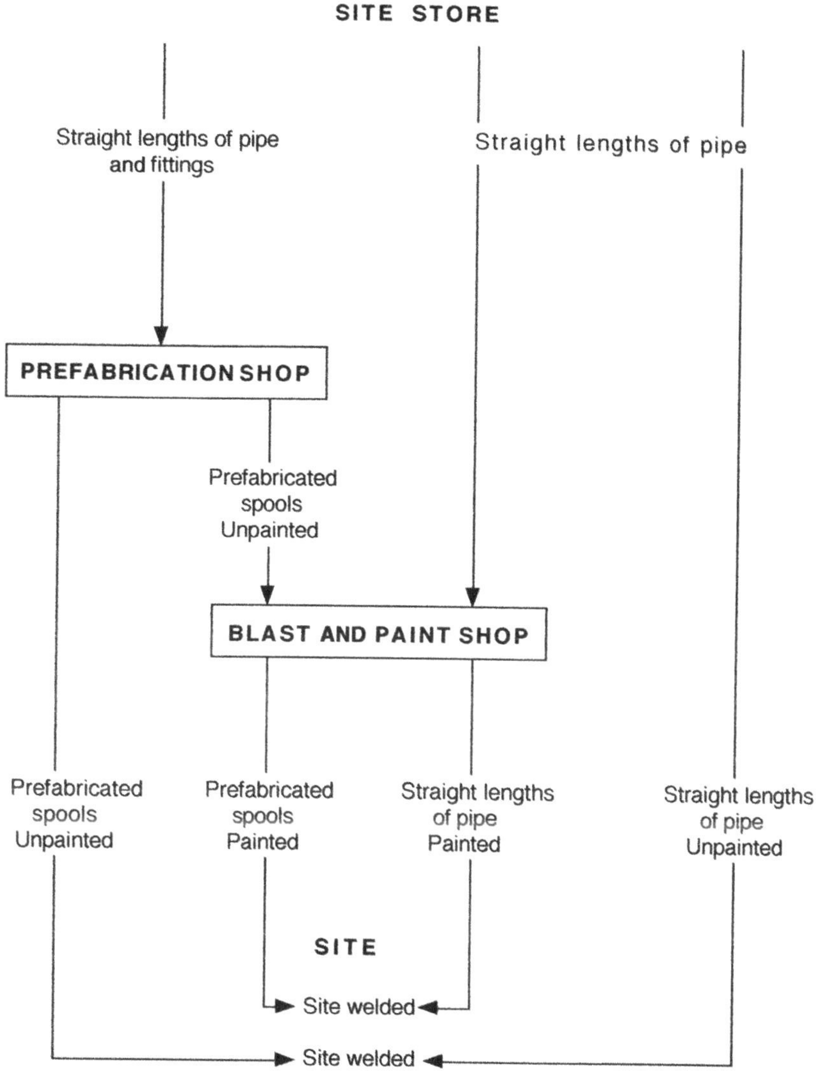

Figure F2 Piping materials – paths taken on site.

needs to be adjusted such that minimal stress is transferred when the bolts are tightened (see Appendix B).

The centre or crossroads of a process plants piping systems is a pipe way at ground level or a pipe rack when the pipes are carried overhead. As part of the construction strategy, installing piping on the pipeway and/or pipe rack may be completed early. It comprises a large proportion of straight lengths. The pipework thus installed is the starting point for taking off connections to items of equipment. The configuration of small diameter piping, may not be predefined on an ISO. Instead its run will be decided on site and welding and erection will proceed accordingly. A range of pipe supports are needed to take the weight of the full pipe, to counteract thrust, to guide it under thermal expansion or contraction and to restrain it when subject to dynamic loads, occasioned by surge, vibration or other phenomena. Supports are also positioned to allow for the weight of isolating and control valves with their actuators, as well as items such as flow meters. Care must be taken during the installation of these in-line items that they have been installed correctly in relation to the direction of flow of the fluid. Attention must be focussed on ensuring that the specified type of gaskets and bolting sets are used for pipework erection and that connections have been torqued to specified values.

Once piping erection is complete there will be a series of procedures to be carried out before it is considered ready for turnover. The ensemble of piping must be divided into **test systems**, which do not necessarily coincide exactly with functional systems, but every element of piping in the plant must be in one test system or another. In fact, a test system comprises piping which will be tested together on the same occasion. Test systems can be defined by the head office engineering department or on site, but whoever does it, the sooner it is done the better. Limits can be defined by mark-ups on P&IDs. These drawings annotated, numbered and filed to represent test systems become master documents to monitor and control the various stages through to completion. When the pressure test is hydraulic these phases are:

- Inspection for completeness of erection
- Flushing

- Pressure testing
- Air blowing and reinstatement
- Release for painting, heat tracing and insulation

Referring again to Figure F1, this sequence is shown schematically, with a variant for pneumatic testing.

Once piping is erected it has to be inspected to verify that it is physically complete. Inspection is carried out by following piping under consideration from one extremity through to the other extremity(ies), meticulously comparing the physical reality with the drawings and reviewing existing quality records. It is important that the persons inspecting are experienced and mature enough not only to spot discrepancies between the works and the specifications/drawings, but also to identify situations and configurations, which whilst conforming to what is specified, in fact are undesirable or detrimental to safe and efficient operation of the plant. These could typically concern access to a valve or insufficient clearance over a walkway. Given the mass of detail to be managed, a preliminary inspection may well bring to light errors and/or incomplete work, which will have to be corrected before the record form can be completed and signed. This intermediate stage can be followed through and closed out using the list of outstanding items proposed in Figure 11.2.

Flushing consists of a throughput of water of sufficient pressure, velocity and duration to remove any dirt and debris which has accumulated during construction. Obviously this task is minimized if care has been taken to close pipe ends temporarily with plastic caps and to encourage welders and pipe fitters not to use an open end as a handy receptacle for a box of welding electrodes or a spanner, which can be easily forgotten.

Preliminary preparations for flushing involve setting limits by fitting blind flanges, creating temporary openings and installing pumps and filters. Items such as control valves, and in-line flow meters must be blinded off, or removed, and where necessary replaced by temporary pup pieces. Some items of mechanical equipment must be isolated to avoid damage and so as not to become receptacles for debris.

The latter may be removed by means of a free discharge or by a filter in the case of a closed circuit. The water used must be of suitable quality and treated with additives according to the nature of the materials present.

It is usual for flushing to occur before pressure testing; the two operations need to be coordinated. Reinstatement will only be made after both are completed. In any case both flushing and pressure testing need to be the subject of written procedures, with marked-up drawings showing the configurations. In those piping systems where water cannot be admitted, cleaning is carried out by means of an air blast rather than by water flushing.

For a given test system it will be necessary to define the test pressure, the duration and the test fluid. Where the test includes both piping and items of equipment, specified test pressures of both must be reviewed before deciding on the value to be chosen for the test. The majority of tests will be hydrostatic with suitably treated water. However, there will be instances where the entry of water will not be allowed, in air instrument piping for example. In these cases the test will be pneumatic, usually with air. The procedure followed must incorporate the necessary safety precautions, with prior formal clearance by the site safety department, to take account of the energy that would be released in case of rupture. Pressure testing is normally a hold point.

Prior to and during the test care has to be taken to ensure that:

- Pressure gauges are calibrated.
- Pressure is raised slowly in steps punctuated by pauses.
- Suitable air purges and drains are in place to ensure no air pockets during the test and no water remains trapped when the system is drained after the test.
- During the hold pressure phase the pressure gauges are monitored to detect any change in pressure. (Exposure of the piping to a hot sun can falsify the readings.)
- Welds, screwed connections and between flanges should be scrutinized to detects any leaks. (For pneumatic testing, this can be done by applying a soap solution, but for reasons of security only permitted at a very low predefined pressure.)

Once the pressure test has been satisfactorily completed and recorded, pressure needs to be released in a controlled manner. Any temporary arrangements can be dismantled and the system blown through with air to remove residual moisture. Some lines may require a chemical treatment, which shall be the subject of a specific procedure. The test system now has to be restored to its pre-test configuration. All items removed need to be reinstated to return the piping system to its initial completely erected status. Once the piping is reinstated, it is formally released for one or more of the subsequent operations of painting, trace heating and insulation.

Appendix G: Electrical Installation

Installation and connection of electrical equipment can only progress once the civil engineering works, buildings and steel structures are substantially complete and mechanical equipment requiring an electricity supply has been erected. Major electrical equipment items concerned include:

- Transformers
- Switchgear and motor control centres
- Batteries and battery chargers
- Electrical elements included in mechanical ensembles, in particular electric motors

These items need to be interconnected with power and control cabling and connected to the earthing system(s). There will be in addition the works associated with lighting, socket outlets and other low power installations.

Besides the EPCM contractor and the owner, the following parties are involved:

- The utility company supplying electricity
- The electrical construction contractor(s)
- Suppliers of electrical equipment
- Suppliers of mechanical ensembles incorporating electrical equipment

A process plant will generally receive its electricity supply from a utility company, but some plants may include on site a significant generating capacity to meet at least part of the load, with the possibility of running in parallel with the exterior supply. This occurs especially when the process generates heat that would otherwise be wasted. The provision of on-site generation can also be dictated by the plant location, where the public electricity supply might be insufficient in capacity, unreliable or even non-existent. It is important that a dialogue be established on site in a timely manner between the owner, the EPCM contractor and the utility company, so that any problem areas are identified and cleared in good time to avoid delays in commissioning.

A very small process plant may be supplied at low voltage, for example 415/240 V three phase and neutral, fed directly from an adjacent substation of a utility company. With a larger load the incoming supply will be at medium voltage, say 11 000 V three phase, feeding a single substation transforming down to low voltage. The largest process plants may receive their electricity at high voltage, via a main incoming substation, from which power is distributed at medium voltage to a number of substations situated throughout the plant. In each of these substations the medium voltage is transformed down to the utilization voltage, 415/240 V three phase and neutral, for example, and possibly also at a higher voltage for the larger motors.

Where essential services must be maintained during a mains power failure, an emergency diesel alternator set can be included in the plant. Once such a set is installed, the shaft alignment checked and all connections made, protective fluids must be drained and systems rinsed. Then lubricating oil, cooling water and fuel can be introduced into their respective systems. Insulation resistance and the functioning of the electrical circuits needs to be verified and the battery charged. The set can then be started manually to enable general operation to be checked, including engine protection arrangements. Then at the first opportunity a test of automatic start-up can be made, with a full load running test.

The majority of transformers installed in process plants are filled with mineral oil, and so we shall limit our discussion to this latter type. Smaller transformers may be delivered sealed and complete; in which case installation is relatively simple. However, for larger transformers, in order to facilitate transport, lifting and handling, they may be drained of their oil, shipped with conservator, bushings and radiators removed, to be reinstated once on site. The oil shipped in sealed drums would be reintroduced into the tank on site in accordance with the supplier's instructions. The tank may be pressurized with inert gas to prevent moisture ingression and oxidation. Quality concerns are essentially:

- Reassembly, joints, gaskets and bolting up
- Name/data plate against specification
- Dielectric strength and moisture content check on an oil sample
- Insulation resistance primary/earth, secondary/earth and primary/secondary
- Winding resistance measurements
- Voltage ratios, voltage vector group, phase rotation
- Tap change arrangements
- Operation of measurement, alarm and protection devices and circuits
- Earthing and cable connections
- Installation and testing of fire protection system (if any)

These points need to be defined in detail in a written procedure, with inspection and test results entered on a complete set of record forms.

Metal clad switchboards comprise an assembly of modular steel cubicles housing circuit breakers, fuses, contactors, bus bars, cable boxes, current and potential transformers, instruments, relays, and so on, with the interconnections. Very small switchboards may be shipped in one piece, but normally they will be broken down into subassemblies to be re-erected on site. Critical operations requiring close surveillance, with appropriate inspection and test points can include:

- Positioning subassemblies, bolting steel frames together, levelling, grouting and tightening anchor bolts
- Connecting steel housing to earthing system
- Interconnecting bus bars, torquing bolts to manufacturer's instructions
- Reinstalling items, potential transformers for example, removed for transport
- Reconnecting auxiliary wiring between modules
- Reintroducing circuit breakers where applicable

- Checking throughout the assembled switchgear or motor control centre for completeness and conformity to approved drawings, including name/data plates, types and ratings of equipment with corresponding identification labels both inside and on the front panel
- Applying insulation resistance and high voltage tests to power circuits
- For auxiliary circuits measuring insulation resistance
- Energizing the auxiliary circuits and testing or simulating all functions
- Making primary and secondary injection tests
- Where applicable racking circuit breakers through service, withdrawn and earthed positions checking isolating arrangements, interlocks and protective blinds
- Verifying cable connections to transformers upstream and downstream, as well as auxiliary wiring for measurement and control circuits
- Carrying out system checks, for phase rotation, interlocks, isolation, protection and metering arrangements before energizing main circuits

The above needs to be developed into a written procedure, with inspection and test results entered on a set of record forms.

Independent battery based sources of power are require for operating switchgear, that is to say, to energize relays and pilot lamps and for circuit breaker closing and tripping, as well as for emergency lighting and for maintaining various critical systems in operation during mains power failure. Each such system comprises a battery, a charger and a distribution board with control and monitoring circuits. The most common types of battery are the lead acid Planté and nickel cadmium types, nominal battery voltages being typically 24, 48, 110, 125 or 220 V. Care must be taken in installing the cells and their interconnection. After the charger has been verified and connections made, the battery can receive its initial charge, a critical operation to be carried out strictly in accordance with the manufacturer's instructions.

In a process plant there are numerous applications for electric motors, in particular for driving pumps, fans and compressors, but also as an integral part of mechanical items of equipment for material handling for example. Where possible these are specified as squirrel cage induction motors with direct-on-line starting for low first cost, ruggedness and simplicity. More complex motor/control ensembles are specified to respond to various exigencies including limiting starting currents, on-load starting against heavy torque, to provide a variable speed drive, or for power factor correction. For some highly critical duties, DC motor drives may be supplied from a battery in case of power failure. Most motors will be fed at low voltage, but with higher power outputs, there can be a technical and economic case for operating at a higher voltage such as 3 300 or 6 600 V. Once the equipment is installed, the motor/driven machine shafts will be checked for alignment by the mechanical construction contractor and the coupling fitted and greased. Insulation resistance of the windings must be measured and protective devices for temperature and vibration checked. With cable and earthing connected and checked, the motor can be energized, with the coupling disconnected to check the direction of rotation.

Induction motors give rise to lagging power factor, which can be penalized in the tariffs applied by the utility company. Certain other items of industrial equipment can also contribute to low power factor. This is compensated by connection of capacitors, either individually for each reactive load, or grouped at a common feed point in the network with banks of capacitors being connected and disconnected automatically in stages as required. Other possibilities for power factor correction include the use of synchronous motors for large drives.

Cables above ground, either in the open or within a building, need to be supported and protected during their entire length by means of cable racks, ladders, trays, conduit, trunking or ducts depending on the type of cable, the physical installation conditions and the degree of protection required. For major cable runs, drawings will define what is wanted, whilst for minor cabling, details may be left to the initiative of those carrying out the work on site. In a given zone, the cable support and protection arrangements have to be completely installed, finished and inspected for formal release before cable pulling begins. It is very important that the structures be sufficiently rigid to support the weight of the cables, that there is adequate space left for any future additions and that the configuration, in particular at changes in direction, takes account of the permitted minimum bending radii of the cables to be installed. Where there are openings through floors and walls, it is recommended to err on the side of largeness in size to facilitate the passage of cables and the carrying out of any alterations or additions. They can always be filled in afterwards if found to be excessive. Good workmanship is important, in particular to avoid any sharp edges or other irregularities likely to damage the cables.

A wide range of cable types is available to meet the various applications. Copper or aluminium conductors can be specified, each type being terminated and connected with a distinctive set of accessories. Concerning insulation, there is a choice of material available to meet every situation. PVC insulation is still specified for the lower voltage and current ratings, but the lion's share goes to cross-linked polyethylene (XLPE) insulated cables. Multi-core cables, especially those in an exposed situation with a risk of mechanical damage, may be protected with single or double layer steel wire armour (SWA). Regarding cables to be installed in aggressive environments, a number of protective outer sheathing materials are available. For cables installed underground, they can be buried or drawn into previously laid ducts, the latter option being useful for road crossings. In both cases coordination with the concurrent civil works is important to avoid interference.

From the quality viewpoint the following areas need attention:

- Close control over the drums of cable, with special attention to identification and sealing the ends to prevent ingress of moisture
- Avoiding excessive tension, friction or abrasion during pulling by careful positioning of guides and rollers
- For buried cables use of sand or sieved earth as bedding and backfill suitably compacted, topped with protective tiles or warning mesh in case of future excavation
- Identifying and protecting buried cables at point of surfacing

- Attaching noncorroding identification labels at regular intervals
- Segregating between power cables and control/ instrumentation cables
- Above ground, fixing and supporting cables at specified interval, respecting bending radii and cable grouping allowed
- High-voltage testing where specified
- Making off cable ends, tightening cable glands and identifying cores and terminals
- Connecting to earth of SWA and other screenings to specified requirements
- Testing for insulation resistance and continuity

Concerning lighting installations in a process plant, the most common choice will be between lighting fittings of the incandescent, fluorescent, mercury or high-pressure sodium vapour types. Lighting circuits as well as those feeding socket outlets are fed from local switchboards. Small transformers may be incorporated to feed extra-low-voltage sockets. For connecting these low power circuits we speak in terms of 'wiring' rather than 'cabling'. A range of wiring systems is available, including multi-core PVC insulated run on cable tray, single core or multi-core PVC insulated enclosed in conduit or trunking or possibly mineral insulated. Besides the inspection and testing of the electrical installation as such, measurements can be taken of the luminous intensity over the lighted area.

A number of different arrangements for a supply system can be specified concerning the relationship between neutral and earth. A discussion on these various options is beyond the scope of this handbook. Instead we shall simply evoke the basic arrangement of separate protective conductors, a main earth terminal in the substation directly connected to buried electrodes. Protective conductors connect normally non-conducting metalwork, such as electric motor frames, switchgear panel metalwork, metal cable trunking and conduits, mechanical equipment and structural steelwork back to the substation main earth terminal. Protective conductors can be dedicated single core cables or copper tapes, or one of the cores of a multi-core cable, usually identified by a yellow and green sheathing. Protective conductors can also be constituted by cable sheathing or armouring. In addition adjacent metallic items of equipment simultaneously within reach of a person may be bonded together by an equipotential conductor to reduce potential difference and hence risk of shock under fault conditions. The substation main earth terminal is connected to the earth electrodes via the earthing conductor. A number of different types of earth electrodes are available including driven rods, buried plates, pipes, tapes or cables, electrodes embedded in concrete foundations, or by using the reinforcing bars of the foundation. Bolted links shall be provided so that the various elements of the system can be separated for testing. A typical test programme for an earthing system could include:

- For the protective conductors and the earthing conductors, witnessing thermal and compression connections, inspecting routing, identification, fixing and protection, checking cross sections and measuring resistances
- For electrodes, checking type, material, dimensions, measuring resistance to earth

Depending on the protective system specified it may be also necessary to execute earth fault loop impedance tests.

In those areas of a process plant where a potentially explosive atmosphere can be present, special measures must be taken in the design of the plant, in the selection of electrical materiel and in its site installation. At the design stage the hazardous areas are classified into zones in accordance with the likelihood of an explosive atmosphere being formed. The response to these risks is:

- Where possible to exclude electrical materiel from the zone at risk
- To use motors, push buttons, lighting fittings, and so forth, specially designed to prevent the coming together under normal, fault or damaged conditions of an explosive mixture

and a source of ignition by sealing the enclosures, by encapsulation or by internal pressurization
- Alternatively, where appropriate, to use equipment so designed that any explosion is contained within and cannot propagate to the exterior

The identity of such equipment is indicated by distinctive markings on the name/data plate and on the apparatus itself tied back to documentation, which in turn is linked to type tests previously carried out on identical items of equipment by an independent recognized authority. The subject of protection against explosion hazards is vast and in most countries is placed within a framework of draconian regulations and directives; so it is essential to meet the exigencies of the authority concerned. Electrical installation work in such areas must in consequence be governed by the strict application of specified requirements and a comprehensive programme of inspection and tests overseen by persons with a detailed knowledge of this field.

Appendix H: Instrumentation and Control

Instrumentation and control (I&C) systems in a process plant cover *inter alia* the processes themselves, equipment operation and protection, emergency shutdown, interlocks, fire and gas detection and protection as well as providing a management information function. In recent years enormous strides have been made and are ongoing in this discipline, especially in terms of data acquisition, transmission, display and storage, as well as in the domain of computer control and process optimization. Treatment of these developments is beyond the scope of this handbook. Instead we shall limit the discussion to basic quality concerns for materiel installation and preliminary inspection and testing. We can classify these activities into one or other of the following categories:

- Initial pre-installation inspection, adjustment, calibration and testing of the various items of materiel delivered to site
- On site installation, inspection and testing of materiel out in the process units and in the utility and offsite areas
- Inspection and testing of I&C materiel delivered to site as part of a packaged unit, such as a motor/compressor set for example, as well as connecting and integrating this materiel into the wider site systems
- Installation, inspection and testing of materiel in and adjacent to the control room(s)
- Installation, inspection and testing of I&C piping
- Installation, inspection and testing of interconnecting cabling including earthing
- Loop testing.

I&C systems for process plants depend for their safe and continuous operation, under normal and abnormal conditions, on independent dedicated electrical, pneumatic and possibly hydraulic sources of power. These power sources should be installed, inspected, tested and made operational as early as practicable, so as to be available for the inspection, testing, calibration and adjustment of the components and ensembles constituting the plant's I&C systems. Batteries provide DC power directly or provide secure AC power through an inverter. Instrument air is produced by oil free compressor sets delivering to an air receiver via filters and air dryers. In an application where hydraulic power is required, the necessary pump sets, filters, reservoir and accumulator, can be integrated into a self-contained power pack.

Section 7.5 evokes site incoming inspection. For all I&C items this means in the first place verification against reference documents, checking name/data plate details, dimensions, configurations, hazardous zone classification and interfaces, that is to say, electrical terminals, cable glands, flanges, screw connections and fixing arrangements. The I&C construction contractor should provide on site a calibration and test facility comprising an enclosed space for the mechanical test rigs for valves and a separate clean zone for instruments. The test apparatus must be of an appropriate level of accuracy, duly calibrated and subjected to regular recalibration. Loose items of materiel delivered to site, such as instruments, sensors, transducers, pneumatic, hydraulic or electrically actuated control valves and also safety relief valves pass through this facility to be tested, adjusted, calibrated and where appropriate lubricated. When I&C materiel forms part of a package, verification can be done *in situ* using portable test equipment. It is necessary to identify and set aside certain factory calibrated items, which normally will not be adjusted on site. All inspection and test results should be entered on appropriate record forms. The test status of each item can be shown by attaching a weather resistant label marked 'bench tested' with the date and signature. Other labels, as part of a formal colour-coded system, can be attached at the time of later changes in status as the plant progresses towards construction completion and turnover and beyond to start-up and commissioning.

Items such as control valves with their actuators and accessories, orifice plates, venturi or magnetic flow meters will be physically installed by the piping construction contractor. However, the I&C construction contractor must closely monitor the operation to ensure that the materiel is handled with due care and attention, avoiding shocks and the ingress of dirt or moisture. These items must be installed the right way round with respect to the indicated direction of fluid flow. They need to be properly supported and constrained, generally not relying just on the piping either side. Before installing an item of I&C equipment, a review should be made of its immediate physical environment, that is to say, adjacent piping, machines, walkways, ladders, handrailing, and so on to detect any potential inconveniences not foreseen at the design stage, which could be corrected. Is there sufficient space? Is the item properly supported and protected? Is the interface with equipment and with piping satisfactory? Could the item be advantageously re-orientated to improve operability or access for maintenance? During pressure testing, flushing and blowing through of the piping, I&C items must, where applicable, be removed and replaced by pup pieces, blanked off or have internals removed to prevent damage or accumulation of debris (see Appendix F).

Adjacent to the above-cited control valves and primary sensors, and at other strategic points, there may be installed in a cabinet or on a chassis, such items as transducers, transmitters, controllers, indicators, push buttons, switches, isolating valves, cable terminal blocks, and so forth, with their associated interconnections, incoming electric and pneumatic power supplies, and connections back to the control room. These need close supervision during installation and a programme of inspection and tests to be defined, implemented and recorded.

Instrumentation and control elements of each packaged unit, a compressor set for example, need to be inspected. There are the protective and operating systems internal to the equipment in the package to be reviewed, then there is the question of compatibility of the I&C aspects of the package with the external systems with which they interface. This should have been taken into account by the EPCM contractor's engineering department during the design and procurement stages, and then checked during shop inspection at the supplier's works before shipment, but sometimes these upstream verifications are not always as thorough as they ought to be or there may have been system modifications which have not been taken into account. For these reasons an early verification, that is, as soon as the package arrives on site, is a wise precaution. A supplier's representative should be available during the more critical phases of installation, inspection and testing.

Installation of sensitive equipment in control rooms and in computer rooms can only commence once the building is finished, dried out, cleaned and painted with its utilities, in particular its air-conditioning system, being fully operational. Making these buildings available early in the overall planning schedule will enable installation to be completed and allow more time for both hardware and software to be tested and developed, in parallel with construction in the process units. Free standing panels, cabinets and consoles, input/output racks, operator stations and peripherals will be unpacked, positioned, levelled, bolted down, earthed, connected to power supplies, then interconnected. Interconnecting cabling needs to be installed, inspected and tested, for insulation resistance, continuity, cable terminal markings and identifications. If the terminals for the cabling out to the field are physically separate and can be easily isolated from the control room equipment, installation and testing of field cabling can proceed in parallel with and independent of operations concerning the materiel in the control room.

A well-defined and documented programme of testing on the computer system supplier's premises before shipment to site of all items connected together or simulated, carried out in the presence of the other involved parties will do much to eliminate any anomalies. However, this does not reduce in any way the need for both hardware and software to be subjected to extensive and detailed verification and development on site. It is vital that for both shop and site testing, procedures, format of quality records, basis of interpretation and criteria for acceptance as well as the responsibilities of the various parties to be fully defined well in advance. Modifications to hardware and software should be logged and be subject to formal authorization. The most important success factor is perhaps the composition of the team drawn from the various interested parties, their motivation and their ability to work together towards a successful conclusion.

Instrumentation and control piping/tubing covers instrument air supply, signal/transmission lines as well as process impulse lines. From the air receiver vessel mentioned above, instrument air is distributed throughout the site to its points of use through a piping system normally prefabricated and erected as part of the piping contract, but from the air headers onwards, small bore piping and the associated isolating valves, filters, pressure regulators, and so on, are normally the responsibility of the I&C construction contractor. For signal/transmission lines tubing used can be of a number of different materials, with the corresponding sets of fittings. Installation and connection calls for meticulous workmanship and attention to detail. Completed piping/tubing needs to be subjected to a pneumatic pressure test.

A process impulse line is the connection between a process fluid contained in piping or in a vessel and a measuring or control device. Therefore the materials used must be compatible with the resulting conditions of pressure and temperature and the nature of the fluid. Special attention is required concerning the configuration of the piping. When the process line is being flushed and pressure tested, precautions must be taken to isolate the instrument. The impulse line should be flushed and pressure tested separately.

I&C cables must be physically segregated from power cables, in particular close parallel runs must be avoided and crossings should be at right angles. Furthermore segregation may be required between I&C cables of different categories. The principles set out in Appendix G for the installation, support and protection of power cables are to a large extent valid for I&C cables. Cables may be multi-core or multi-twisted pairs. According to application each pair may be individually screened or a single screen may be included for the whole cable. Depending on the degree of protection required against mechanical damage, cables may incorporate armouring and protective sheathings. The inspection and test programme must include insulation resistance and continuity. In addition some cables may require measurement of inductance, capacitance and loop impedance. Inspection of cable terminations covers cable glands, core and terminal identification, screening and armouring. Extra length of cable can be coiled adjacent to the cable entry to facilitate removal of the equipment, to facilitate reconnection and possibly to take account of any vibration. Coaxial cables should be tested in accordance with the manufacturer's instructions.

Earthing is not only necessary for the safety of personnel and for the protection of materiel in case of an earth fault occurring, but also contributes to the efficient operation of the systems. Depending on the particular application, connection may be made to the plant earth or to a separate earthing system with its own buried electrodes, or even to more than one system. Earthing of screening and armouring must only be made at specified locations. To achieve this certain gland plates will be insulated.

Concerning fibreglass optic cables, the whole procedure of installation, cable termination and jointing, as well as subsequent testing needs to be carried out strictly to the manufacturer's requirements. After installation continuity and attenuation are verified.

In those areas of a process plant where there is a risk of potentially dangerous explosive atmosphere being present, the concerned systems must be designed and materiel specified to meet these conditions. Equipment bear the appropriate markings, are accompanied by and traceable to official documentation acceptable to the relevant authorities. This necessitates close incoming inspection of materiel, supervision of erection by competent supervisors and rigorous compliance with specified requirements (see Appendix G).

A loop is a combination of a number of items of equipment interconnected to measure and/or control a process variable. A list of loops, with the definition of their contents, interfaces and limits is prepared as part of the engineering design process. A prerequisite to functional testing of the loop is that:

- All items have been inspected, calibrated and tested before installation

- All items have been inspected, adjusted and tested after installation as may be necessary
- All interconnecting cables are in place, duly connected, inspected and tested
- All pneumatic and hydraulic piping is in place, duly connected, inspected and tested
- Electric power supplies are connected and checked
- Instrument air supply is connected and checked
- All the above has been duly documented and reviewed
- A test procedure is available including means to record test results
- All necessary test and simulation equipment is available
- A team is formed from the various parties present and the responsibilities of each participant defined

Testing comprises essentially applying signals at the primary detecting elements, either directly or by simulation, then tracing the sequence through transducers, inputs/outputs, interlocks, control and logic circuitry, local/remote displays, to the indicating or controlling element.

Appendix I: Painting

The EPCM contractor will have developed a project paint specification, generally in collaboration with one or more paint manufacturers, who may offer technical assistance and possibly a guarantee arrangement. The specification will include a range of paint systems to cover project requirements for the different situations to be found in the future plant under operating conditions, that is, outdoor/indoors, dry/humid, corrosive/noncorrosive, high/low temperature, all against the background of the site prevailing climatic conditions. Whilst paint may be applied to other materials, this appendix is focussed on painting of steel surfaces. The following cases will occur on site:

- Items arriving on site with at least a primer coat of a paint system already applied, the subsequent coat(s) being added *in situ* after erection
- Subassemblies, notably in the piping and structural steel work disciplines, possibly prefabricated on or adjacent to the site, to be blast cleaned and receiving at least the primer coat before erection, the subsequent coats being applied *in situ* after completion of erection
- Items such as site-erected reservoirs, comprising prepared plates and prefabricated items, welded together on site, to be blast cleaned and to receive all the coats of the paint system after completion.

If a steel surface is to be protected on a long-term basis, three conditions must be met:

- The surface must be properly prepared.
- An appropriate paint system must be used.
- The different paint coats must be properly applied.

Problems which occur subsequently can nearly always be traced back to the non-respect of one or more of these precepts.

Surface preparation and painting should only take place when measured surface temperature and relative humidity are within specified limits, applying a margin above dew point. The necessary measuring instruments and data must be available in the location where the work takes place. Measurements should be taken and logged at regular intervals during a shift as a standard procedure, special attention being exercised during changeable weather conditions. If the specified limiting conditions are approached, work must stop.

Hand- or power-wire-brushing may be specified in some circumstances. However, the vast majority of cleaning operation on or adjacent to the site will be carried out by means of projected abrasives. The objective is to remove rust and mill scale and to roughen the exposed bare metal surface to create an anchor pattern to enhance adhesion of the primer coat to the metal surface. A range of abrasives is available, either disposable or able to be recycled. It is important that dirt, oil and grease be removed from the metal surface before blast cleaning. Standards define a number of levels of surface preparation in the form of photographs. Inspection is based on the comparison of the surface with the photo of the specified level. Instruments are available to give an indication of roughness profile. Proper blast cleaning followed by the timely application of the primer coat is the most critical operation, so a hold point is normally fixed between these two activities.

A large range of paints types is available to respond to every surface protection need. Trade names do not always reveal the family to which the paint belongs, but full details can be found in manufacturers' data sheets. There are two broad classifications concerning drying and curing:

- Air drying paints, which depend solely on the evaporation of solvents and thinners
- Paints, whose curing is provoked by chemical reaction, including two-pack paints

Paint tins must be stored under cover within the temperature limits set by the manufacturer and used during the indicated shelf life. For two-pack paints the products must be brought together to form a consistent mix, to be applied without delay within the pot life limit defined by the paint manufacturer.

A paint system is defined as:

- A grade of surface preparation, that is, cleanliness and roughness profile
- Mutually compatible and complementary paints to be used for the different coats with colours specified
- Dry film thickness (DFT) for each coat with an indication of the corresponding wet film thickness (WFT)
- Time limits between the successive operations of blast cleaning and the application of the primer, then between successive coats

For a given item of equipment or class of material, the paint system to be applied will be defined in a specification, on a data sheet or on a drawing.

Painting is a domain in which care and attention at every stage of the process pay dividends, and where negligence will sooner or later lead to deteriorating surface protection. The spraying equipment must be thoroughly inspected before use. In addition the paint to be used must be checked for conformity with the specified system and to see that shelf life has not been exceeded. Assuming that the surfaces have been blasted to an acceptable standard, the primer must be applied as soon as possible and certainly not later that the specified time limit. Whilst paint can be applied with a brush or roller, most painting will be done by spraying, either by low pressure air spraying with precautions taken to render the air free of oil and moisture, or by high pressure airless spraying. Successful paint application depends on operator skill, which can be consolidated by competent supervision.

For each coat, WFT can be checked with a gauge during application, the value specified being an indication related to the corresponding DFT. The latter can be measured with a meter previously calibrated, which depending on the model may be able to record and print out a number of readings and calculate an average value. For the second and subsequent coats, DFT measured will be cumulative. Depending on the manner in which DFT is specified, some interpretation of the measured values may be necessary. A visual inspection between each coat can detect defects, with any repairs being executed before the next coat is applied. If specified, adhesion, holiday and pinhole tests can be executed.

Blast cleaning and painting may be in progress adjacent to each other. Ensure that the first does not contaminate the second and that after blasting the surface is brushed and vacuum cleaned before the primer coat is applied. Cleaning a painted surface before the application of the next coat is also important.

To sum up, quality of the finished painting is dependent on the following:

- Assigning a competent and proactive supervisor
- Verification of equipment and materials before use
- Working within limits of ambient conditions
- Respecting timing requirements, in particular for application of primer following blast cleaning, and between successive coats
- Blast cleaning and applying paint to good trade practice
- Ensuring DFTs attained are within specified limits, using WFTs as a preliminary indication
- Identifying defects and correcting them before they are covered up
- Defining and implementing a formal inspection and test programme, with results duly recorded in writing.

Appendix J: Thermal Insulation

Thermal insulation may be specified for items of equipment and for piping systems for hot or cold service, to conserve energy, for process reasons or to prevent freezing, but also for personnel protection. There will be a specification produced by the EPCM contractor comprising a number of insulation systems to cover the needs of the project. For each system the following will be defined:

- Type of insulating material used, in rigid or blanket format
- Thickness and number of layers (joints to be staggered)
- Type of banding
- Material and thickness of protective jacketing
- Details of jacket edge swaging, screw fastening and sealant

The specific insulation system to be applied to a particular item of equipment or pipe line will be defined on a data sheet, drawing, or equipment or line list. Apart from the need for a technically sound job, the finished insulation should be neat and pleasing to the eye, for insulation is highly visible. The material must be stored under cover and insulation work must not be carried out when exposed to rain or snow.

Insulation can only be put in hand if it is sure that all previous steps have been completed. Depending on the particular item of equipment or line, these can include:

- Flushing, pressure testing, blowing through, reinstatement, and over all inspection for completion
- Painting
- Steam or electrical heat tracing

These stages must be recorded, leading to a formal written release for insulation work to be put in hand. Equipment design must take account of the insulation to be added by the provision of support rings and by dimensioning name/data plate brackets, ladders attachments, nozzles and vessel manholes to allow for insulation thickness. Piping design must also take account of insulation. Before launching an insulation campaign it is a good idea to regard the first few items as prototypes, which can be reviewed and any difficulties resolved before embarking on the main body of work. The special requirements of valves and flanges must be met.

Quality control consists of ongoing surveillance with a formal inspection hold point before each cover-up with a final inspection on completion. The sequence can be summed up as follows:

- Obtain release of equipment or line for insulation
- From documentation determine insulation system to be applied
- Check that equipment or piping design takes account of thickness of insulation
- Verify material for conformity to requirements
- Fit first layer – inspect
- Fit second layer (if there is one) – inspect; joints must be staggered with respect to those of previous layer
- The same again for any subsequent layers
- Fit protective jacketing – final inspection

These steps must be recorded on a suitable form.

Index

Printed and bound by CPI Group (UK) Ltd, Croydon, CR0 4YY

15/09/2026

14962393-0002